Practical Cell Analysis

Practical Cell Analysis

Dimitri Pappas
Dept of Chemistry & Biochemistry, Texas Tech University, USA

A John Wiley and Sons, Ltd, Publication

This edition first published 2010
© 2010 John Wiley & Sons, Ltd

Registered office
John Wiley & Sons Ltd, The Atrium, Southern Gate, Chichester, West Sussex, PO19 8SQ, United Kingdom

For details of our global editorial offices, for customer services and for information about how to apply for
permission to reuse the copyright material in this book please see our website at www.wiley.com.

Library of Congress Cataloging-in-Publication Data

Pappas, Dimitri.
    Practical cell analysis / Dimitri Pappas.
       p. cm.
    Includes bibliographical references and index.
    ISBN 978-0-470-74155-9 (cloth)
  1. Cytology–Technique.   I. Title.
    QH585.P357 2010
    571.6028–dc22
                             2009049247

A catalogue record for this book is available from the British Library.

ISBN 978-0470-74155-9

Set in 10.5/13pt, Sabon by Thomson Digital, Noida, India.
Printed and bound in Great Britain by TJ International, Padstow, Cornwall.

For Mimi, Anya, Natalya, and Micah.

Contents

Preface

This text came about for one good reason. As analytical chemistry and biology move closer together, biologists are performing increasingly sophisticated analytical techniques on cells. At the same time, chemists turn to cells as a relevant and important sample to study using newly developed methods. In both fields, there is a level of knowledge, usually passed down from researcher to researcher, which is not commonly found in the literature. Techniques, hints, and tips that can save time and effort – or avoid artifacts – that are "common knowledge" to one field are often hidden to another. For example, learning flow cytometry is often an art, as the number of adjustable parameters can turn a well-prepared sample into garbage once data acquisition begins. Similarly, developing a microfluidic culture device requires an understanding of the cell biology that dictates cell adhesion, growth, and response to shear stress. Setting up a culture lab, while trivial to a biologist, can be initially viewed as a daunting task by a chemist trained in classical procedures. Conversely, many analytical techniques require an intimate knowledge of how to properly acquire data. An understanding of the analytical principles, and the cell biology, can lead to successful research combining both.

WHY STUDY CELLS?

Research involving biological systems can occur on several levels. Each level of research, from molecule to organism, has distinct advantages and disadvantages, depending on the problem under investigation. The molecular level of bioanalytical research can elucidate interactions between the underlying machinery of a biological process. Molecular analysis, while highly detailed, lacks the *in vivo* mechanisms that often interact on a higher level than the enzyme–substrate (or similar) case. *In vivo* work includes the full interaction of the living system. When looking at the entire organism, particularly a complex one like a mammal, it is difficult at times to separate the response of interest from all of the potential interfering signals and artifacts. Cellular analysis – whether with primary or immortal cells – lies in between the full-fledged organism and its molecular underpinnings. Molecular processes can be studied in living cells, and many observations of living cells can be used to predict the *in vivo* process. In addition, cell research often has fewer restrictions that *in vivo* work (especially if primary cells from one animal will be used for many experiments). In many cases, cells of interest contain most – or all – of the *in vivo* functionality, or can be used to extrapolate response from the entire organism. In the case of blood cells, the response of the organism can be readily determined from the cell sample in most instances. Pancreatic islets, while technically clusters of cells, can be isolated to study the production of insulin for diabetes research. Muscle cell contraction, on the other hand, can be studied on the cellular level, but lacks the anchoring to a physical frame that is found *in vivo*. Therefore, cell research must be conducted judiciously, so that experiments are warranted and can be used to understand organism response.

From the earliest days of cell analysis, it has been a marriage of the tools and methods that has allowed scientists to peer into the cell and unravel its mysteries. From the simplest light microscope to the newest microfluidic device, the ability to analyze the cell as an analyte, and as a container of analytes, has enabled a host of biomedical problems to be studied.

STUDYING CELLS

When faced with a biomedical problem to investigate or solve, the choice of both cell type (the sample) and analytical method are critical. Often, more than one technique will yield comparable information. In other cases, two or more techniques can be used to provide complimentary information. For example, fluorescence microscopy can yield high spatial and temporal resolution images of cell structure and morphology, but with low cell counts. Flow cytometry, in most cases, cannot yield any morphological information. However, the high cell counts and multi-parameter measurements can compliment data obtained by fluorescence microscopy. Cell culture on a microfluidic device can be coupled to fluorescence imaging, or cell separations. In many flow cytometry applications – particularly those involving rare cells – a cell-separation step beforehand can enrich cell concentrations and provide better results.

This book discusses cell analysis from setup of a laboratory for cell work to using specific analytical methods. The goal of this book was to create a practical guide for working with cells in an analytical instrumentation setting. Therefore, Chapter 1 deals with acquiring cells, cell types, and how to choose a cell line or primary cell. Chapter 2 discusses the cell laboratory itself, from sterile handling equipment, incubators, and common items found in a cell lab. Floor plans of two laboratories serve as examples of the ergonomics to consider when working with cultures in a sterile manner. Chapter 3 discusses culture medium, additives, and the practical aspects of maintaining cells for analysis.

From an analytical standpoint, an understanding of the intricacies of cells can avoid many artifacts. For example, Chapter 4 discusses microscopy (e.g., light transmission, fluorescence, and atomic force) techniques for cells. In the case of fluorescence microscopy, the cell is a fixed object that is subject to photobleaching, toxicity, and loss of viability in long-term imaging. Understanding how to avoid photobleaching, and how to develop a chamber amenable to long-term cell imaging, can enable long-term experiments with high temporal resolution. Techniques to maintain cell viability in microscopy are also critical, especially for biological processes, which can take significantly longer than many chemical reactions (traditional chemists are not concerned with viability). Staining techniques, artifacts when making sensitive fluorescence measurements, and the sacrifice between strong statistics and spatial resolution are all discussed.

Chapter 5 deals primarily with cell separations, including fluorescence-activated cell sorting (FACS). Cell separation techniques are becoming both increasingly popular and diverse. Methods of producing a pure cell sample, based on differences in size, morphology, electrical properties, or antigen expression can be used individually or in tandem. Separations of living cells are both an analytical (i.e., cell isolation and counting) and a preparative method, an enabling technology for other analyses. Whether the separation method involves magnetic particles, droplet sorting, affinity chromatography, or other approaches, the fundamental aspects of cell isolation and reducing false positives are present in every separation strategy. Methods to reduce nonspecific capture, to enrich rare cells, and to combine techniques for greater separation power are all presented.

While FACS separations are discussed from a principle standpoint in Chapter 5, the mechanisms and detection are discussed alongside flow cytometry in Chapter 6. Flow cytometry is one of the earliest cell analysis techniques. While it has matured and evolved over the decades, new methods and instrumentation continue to make this a vibrant field. Flow cytometry is often heralded as an objective technique (relative to microscopy, which can be highly subjective). However, given the number of parameters that must be set for an analysis, it is possible to skew data, or to produce artifacts. Compensation, the effect of detector sensitivity, and multiple occupancies are just some of the obstacles to obtaining suitable data from a flow cytometer. Once a good routine has been established with the instrument, a flow cytometer is then capable of producing a wealth of information from a cell sample.

Microscopy, cell separations, and flow cytometry are some of the most common cell analyses performed around the world. They are, largely, macrofluidic systems requiring large sample volumes and a greater degree of operator intervention. The continuing interest in "lab-on-a-chip" (microfluidic) devices has created a new form of cell analysis, where the fluid scales approach the scale of the cells themselves. Chapter 7 discusses microfluidic fabrication methods and ways to analyze cells by microfluidics. Many of the techniques discussed in preceding chapters can be applied to or integrated with microfluidic devices to increase information content or expand analytical capabilities.

HOW I GOT INTO THIS

My graduate and post-doctoral background are, I must admit, in no way related to cellular analysis. I studied laser excitation of a small cloud of cesium atoms. In fact, I don't recall making a single solution in the 5 years I spent in Jim Winefordner's and Nico Omenetto's laboratories at the University of Florida. What I did learn, aside from some fun and interesting spectroscopy, was the ability to apply analytical thinking to new problems. Therefore, when I left Gainesville, FL, for the equally humid shores of Houston, Texas, I was prepared for my new life as a bioanalytical chemist at NASAs Johnson Space Center. As a contractor with Wyle Life Sciences, I was thrust into a dynamic (and fun) group of people cramped into a lab roughly the size of a small recreational vehicle. I had never seen a cell incubator before, or even a cell since I was in high school biology class. Immersion is the best learning strategy, and within a week I was feeding my own, sterile culture of baby hamster kidney cells, the weed cell of our lab. It was during those few years at NASA that I realized two very important things. First, cell analysis – setting up a lab, maintaining cultures, handling cells – was not as difficult as first perceived. The second thing I noticed – and this is no slight to my biologically inclined colleagues – is that biology and chemistry are often quite different things, despite our best efforts to integrate the two. Biologists have a wealth of unwritten knowledge for cell handling and culture, but still like to use gels – those antiquated slabs of acrylamide that are like cavemen's clubs compared to modern electrophoresis methods. There was at times real disconnection between the chemists – whose idea of a clean sample was one that was not turbid – and the biologists. Yet we shared common ground and common problems. This book, therefore, aims to bridge some of those problems and make connections between the two fields. For the analytical chemist, this book is aimed to orient him or her to the cell-culture laboratory, and the practices and considerations of measuring cells. For the biologist, newer – but readily available – technology is discussed to enable new biological analyses.

Rather than list new techniques that may never find commercial or academic fruition, this book is aimed at the practical, and at the readily implemented. Not every reader will have access to two- and multi-photon excitation microscopes, discussed in Chapter 4. However, everyone will be able to construct his or her own perfusion chamber for microscopy, for a minimal financial investment. This book contains numerous figures, flow charts, and tables aimed at deciding which techniques/samples to choose, and how to troubleshoot unforeseen problems as they arise. To keep the book as practical as possible, I have limited theoretical discussion when deemed excessive or unnecessary. It is my hope that this book rests on the laboratory bench (preferably away from the blood-borne pathogens), rather than on a shelf in the lab.

HOW THIS BOOK IS PUT TOGETHER

This book is meant to be a useful, practical guide. Much like a good manual or cookbook, the information should be easy to find. The main chapters (1–7) deal with the fundamentals and applied aspects of each technique. Chapter 8 discusses statistical considerations of analyzing cells. While some protocols are found in their respective chapters, many of the protocols (particularly those that can be applied to more than one technique) are placed in Chapter 9. Chapter 9 also contains several tables of useful probes and standards that can be used in many different cell analyses. Within each chapter, useful hints and tips are emphasized for easy reference. Like any new idea or technique, there is a bit of trial and error, of learning, in the cell-analysis process. This book aims to share some of these lessons and point out pitfalls and obstacles along the way. Cell analysis is an exciting field that truly has limitless possibilities. As new problems arise that can be solved with cells, new analytical techniques are needed. The marriage of cell biology and analytical chemistry is a sensible one, and, with care, that union can help to understand some of the major health problems facing the world today.

Acknowledgments

This book is the product of a year of research and writing. During that time, and in the years leading up to it, several people influenced the material, or were responsible for some of the career turns that led me to start writing this book in the summer of 2009. I will, undoubtedly, have forgotten someone in this list of acknowledgments, but I will start with those who made this book a reality. Jenny Cossham of John Wiley & Sons worked with me from the book's conception to its final publishing. Jenny's initial email was what started this project, and her hard work and constant support were integral to its success. Gemma Valler and Zoe Mills, my production liaisons, were always quick with answers and enthusiasm. I am grateful for my current and former graduate students (Kelong Wang, Sean Burrows, Ke Liu, Randall Reif, Michelle Martinez, Yu Tian, Peng Li, and Yan Liu) and undergraduates (Charmaine Aguas, Ximena Solis-Wever, Brandon Cometti, and Molly Marshall, among others). Their dedication to our research efforts allowed me to focus on this and other projects. My current and former colleagues at Texas Tech made life easy for me while I wrote this book. I must also thank my former colleagues at Wyle Life Sciences and NASA Johnson Space Center, from whom I learned many of the tricks I've shared in this book. Ariel Macatangay, Grace Matthew, Dianne Hammond, and Sarah Wells were instrumental in my introduction to the world of cell analysis. Jim Winefordner, Nico Omenetto, Ben Smith, and my colleagues at the University of Florida taught me how to approach problems with an open mind. I would also like to thank Bob Kennedy of the University of Michigan and Edgar Arriaga of the University of Minnesota for their support of my research career; better advice would have been hard to find. This work was supported by a grant from the Robert A. Welch Foundation (Grant D-1667).

I would finally like to thank my wife, Mimi, for her unflagging support of both this book and my academic career. Her constant editorial guidance and patience made this work possible. Most of this book was written while my children slept at night, and so it is to those pleasant dreams that I dedicate much of this work.

1

Getting Started
(and Getting the Cells)

1.1 INTRODUCTION

In any type of cellular analysis, one must consider both the analytical technique to use, as well as the cell type. Rather than start this text with a discussion of how to set up a cell-analysis lab (Chapter 2) or maintain cultures (Chapter 3), this chapter discusses the practical aspects of obtaining cells, regardless of what analysis is required. There are two possible scenarios in which an analytical scientist encounters cells; either the cells define the problem, or the scientist is in search of cells to validate a technique. In the case of the former, the application drives the cell type. When the pioneers of flow cytometry began their work decades ago, the samples dictated how the instrumentation would take form. The main cell types of interest at the time were blood cells – for both their tremendous health relevance and for their suspension qualities – as well as cells removed for gynecological screening, among others. The need for fast cell measurements (Chapter 6) drove the technology, but the cell samples were ready and waiting for their scientific counterparts.

In other cases, the scientist finds himself or herself with an exciting new technology that may one day change the landscape of cell research in a manner similar to microscopy and flow cytometry. The technique, perhaps first tested with beads or some other cell simulant, now requires "the real thing." Perhaps the scientist has already validated this method with cultured cell lines, and wants to move on to the truly "real thing," primary

Practical Cell Analysis   Dimitri Pappas
© 2010 John Wiley & Sons, Ltd

cells. No matter what the driving factor, the choice of cell type, and the origins of that cell sample are as critical as any aspect of the instrument design or sample-processing protocol. An excellent technique with the wrong cell type (e.g., an antibody–antigen mismatch) or a cell type that fails to attract interest, can fail as assuredly as a technique with poor figures of merit.

This chapter covers the types of cells one may consider for a cell-analysis technique, primary or cultured, animal type, prokaryotic/eukaryotic, and so on. Some example protocols for primary cells are given, as well as sources for cultured cells. More importantly, the little-shared pitfalls of choosing and obtaining cell cultures are also discussed, as well as methods to avoid them. Later chapters in this book cover the analytical techniques; for now, the discussion will be restricted to the cell as a commodity in the laboratory.

1.2   THE DRIVING NEED

Are the cells specific to a biochemical or medical problem? If so, then much of decision work is eliminated, although obtaining the cells may not be much easier. If not, are the cells going to serve as validation of a method? In this case, does the species matter, or the antigen expression? Is cell morphology important? Figure 1.1 depicts some of the questions one may consider while choosing the cell type if a specific species/tissue type are not already defined. For example, if a group has received funding to study smooth muscle cells, then the species may not matter, but the cell type does without question. In this case, the decision of whether to use primary or secondary cells must be made (Section 1.3). Figure 1.1 will guide users who need a cell type to demonstrate a method. For example, if a new cell culture device is developed specifically to accommodate suspended cells (see Chapter 7), then almost any suspended cell will suffice, and therefore one may opt for a well-characterized, immortalized cell line of a particular species, one that can be perpetuated indefinitely in the lab for long-term quality control. In other cases, it may be prudent to use an established cell line that is closely associated with the analytical technique in question.

Certain cell-analysis techniques require some type of reporting capability or response to a stimulus. Cells transfected to express green or other fluorescent proteins are useful in imaging cell growth in culture devices, for detection in cell separations, and so on. Cells transfected for a specific study, such as using cyan and yellow fluorescent proteins attached to two different target proteins, can be developed in one's own laboratory, and

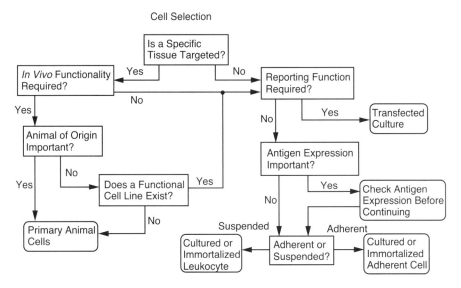

Figure 1.1 Selection of cells based on experimental need and type of analysis. Where *in vivo* functionality is required, primary cells are typically needed, unless a cultured cell line retains the desired phenotype. For cultured cells, antigen expression, morphology, and so on must be considered

become an invaluable resource. Methods for transfection, as well as methods for labeling cells with large and/or impermeant materials are discussed at the end of this chapter.

1.3   PRIMARY AND CULTURED CELLS

When first developing a technique, it is tempting – and often advisable – to begin with a cultured cell line that is well characterized and easy to obtain. However, cultured cells have many drawbacks, which preclude their use in all applications. For example, if a particular organ, tissue, or animal is to be studied, it is more straightforward to obtain cells from that animal and tissue type rather than hunt for a cultured cell line of the equivalent type. Of course, if human cells are of interest, then primary tissue and cells are difficult to obtain, and a cultured line may be the only choice. There are several disadvantages and advantages to primary and cultured cells, which are summarized in Table 1.1.

This discussion will begin with cultured cells. If cultured cells are the chosen route to cell analysis, then the benefits of relatively stable cell behavior, morphology, and growth are apparent immediately. A short

Table 1.1   Comparison of primary and cultured cell lines

	Primary cells	Cultured cells
In Vivo Cell Function Retained	Yes, until differentiation	Most cell lines do not retain *in vivo* functionality
In Vivo Morphology Retained	Yes, until differentiation	Many cell lines (e.g., leukocytes) retain *in vivo* morphology
Change in Protein Expression	No, until differentiation	Yes
Ethical Challenges[a]	Moderate to high	Low
Cost to Obtain[c]	Moderate	Low
Human Cells Readily Available	No, samples are difficult to obtain	Yes, many cell lines are derived from humans
Animal Cells Readily Available	Yes	Yes
Cell Longevity[b]	Cells reach senescence or differentiate after a number of passages	Most cell lines replicate indefinitely (some become senescent after a number of passages)
Immortalization	Primary cell lines are not immortal	Most cultured lines are immortal
Aneuploid	No, unless diseased cells are targeted	Many cell lines are aneuploid, or become so after multiple passages

[a] Cultured cells pose few ethical restrictions, as the initial harvesting has already been conducted. Primary cells will require IACUC approval for animals or IRB approval for human subjects.
[b] Before death, senescence, or differentiation from their original form.
[c] Maintenance costs, after obtaining the cells, are assumed to be equivalent.

survey of recent published literature will reveal that while a vast number of cultured (immortalized) cells are used, there are a handful that are used with a high degree of popularity in cell analysis (Table 1.2). However, one should carefully research which cell types would be best for a given analysis, rather than following the lead of others, or using the same cell lines as another laboratory just to ease the decision-making process.

When selecting a cultured cell, one must decide on what is most important; cell morphology, growth, expression of a reporter, and so on. The first and easiest decision is to narrow down what cell type is to be analyzed. If neuronal activity is of interest, then the cell line PC12 may be a logical choice. PC12 cells differentiate in the presence of nerve growth factor [1], but otherwise remain immortalized, allowing for a permanent culture to be established, but for differentiated cells to be generated when needed. Regardless of cell type, one must also address the differences in functionality between cultured and primary cells. For example, while some cells can be differentiated in culture to take on

Table 1.2 Example cell lines used in bioanalysis

Cell line	Organism	Tissue	Growth type	Morphology	Disease state	Notes
Chinese Hamster Ovary CHO	Chinese Hamster	Ovary	Adherent	Epithelial		
HeLa	Human	Cervix	Adherent	Epithelial	Adenocarcinoma	Aggressive growth properties, high cross-contamination probability
Jurkat	Human	Blood	Suspended	Lymphoblast	T-cell lymphoma	Used in apoptosis studies as a positive control
U-937	Human	Blood	Suspended	Monocyte	Lymphoma	
NIH-3T3	Mouse	Embryo	Adherent	Fibroblast	None	
RBL-1	Rat	Blood	Suspended	Lymphoblast	Leukemia	
HT-29	Human	Colon	Adherent	Epithelial	Colorectal adenocarcinoma	Variable size and morphology (relative to normal epithelial cells)
HL-60	Human	Blood	Suspended	Myeloblastic	Leukemia	
Molt-3/Molt-4	Human	Blood	Suspended	Lymphoblast	Leukemia	Both cell lines come from same patient and express CD4
HuT 78	Human	Cutaneous T cell	Suspended	Lymphoblast	Sezary syndrome	Expresses CD4, variable morphology, clumping behavior
Raji	Human	Blood	Suspended	Lymphoblast	Lymphoma	Expresses CD19, good B cell model
PC-12	Rat	Adrenal Gland	Adherent/ Clustered	Polygonal/ Aggregates	Pheochromo-cytoma	Can differentiate into neuron-like cells
CCEM-CRF	Human	Blood	Suspended	Lymphoblast	Leukemia	Expresses CD3 and CD4
HL-1 (10)	Mouse	Heart	Adherent	Cardiomyocyte	None	Immortal and exhibits contractile behavior. Excellent heart model
C166-GFP	Mouse	Yolk Sac	Adherent	Endothelial	None	Clone of C166 line, expresses GFP
CCD-1064Sk	Human	Skin	Adherent	Fibroblast	None	Senescent after 54 passages, 46 chromosomes, not immortalized
Detroit 532	Human	Skin	Adherent	Fibroblast	Down Syndrome	Senescent after 30 passages, 47 chromosomes

phenotypes similar to primary cells, other cells cannot be made to resemble their primary counterparts. If muscle contraction is required, then there are several cell lines that retain contractile function, even if some of the morphological features are lost.

When considering cultured cells, one must consider the end experiment. If fluorescent protein reporting, or other mutations, are required, then transformation of a primary or cultured cell line is required (Section 1.14). Will adherent or suspended cells be necessary? If one is validating a culture device (Chapter 7), then adherent cells – regardless of type – will be necessary. However, for validation of flow cytometry, MACS, affinity separations, and FACS analyses of suspended cells will make sample preparation easier. Suspended cells, particularly blood cells, are readily available in both primary- and cultured-cell samples. The benefit of suspended cells is that culture (for proliferating cells) is simple, and the ethical issues of extracting primary cells in this manner are fewer than when tissue samples are required.

1.4   CHOOSING A CULTURED CELL

Once the decision to obtain a culture from a transformed cell line has been made, Figure 1.1 can help decide on the specific cell type. First, if reporting functionality is required, a cell line already transformed – such as the mouse endothelial cell line C166-GFP from the American Type Culture Collection – can be used to demonstrate devices or develop analytical methodology before other cell lines are tested. It is also possible, as noted below, to transfect a cell line, provided the researcher has met all guidelines for transforming cells and handling recombinant DNA, and so on. The latter approach allows one to develop many different cell lines into reporter clones, but requires additional infrastructure and cost. If a fluorescent cell is needed, and there is no need to tie the fluorescent protein to the expression of a particular protein, then several cell lines are readily available commercially (See Table 1.2). It is also possible to use a long-term (1–5 days) fluorescent tracer, such as the CellTracker series of dyes from Invitrogen, to render a cell fluorescent for tracking or detection purposes.

Whether or not reporting functionality is needed, there are applications that require cells of a certain tissue or animal type. If this is the case, then one must exercise caution to avoid choosing a cell line that lacks a key feature. For example, not every cancer cell line displays tumorigenic activity if injected into animals or cultured with other tissues. If the cell analysis in question in this example is to study tumorigenesis, then this critical aspect of the cell line must be investigated before a cell line is selected. Similarly, antigen expression is critical for many applications, such as flow cytometry and affinity or MACS cell separations. Many transformed cell lines are derived from cancerous cell lines that have deviated from the original tissue. Antigen expression should therefore be investigated before acquisition or at the least tested by flow cytometry before the cell line is used in additional analyses.

In some cases, morphology (beyond adherent or suspended), species, functionality, and antigen expression are unimportant. Examples of these cases include the development of a new analytical or culture technique, where the cells in question must simply survive long enough to prove the concept will work. In these circumstances, a well-characterized cell line that will grow under a wide range of conditions may be the ideal choice. These so-called "lab weeds" are robust cell lines that are used routinely around the world. Examples such as Chinese Hamster Ovary (CHO), Jurkat, HeLa, and other cell lines are nearly as ubiquitous as the analytical balance or pH meter in modern cell laboratories. These cell types are also available in various transfected clones for a wider range of options. Well-characterized standard cell types are not always the best choice, as they suffer from problems associated with immortalization (see below) and often offer the bottom-line choice. There are thousands of cultured cells from different species, tissues, and phenotypes available. Straying from the standard choices may allow for greater impact in one's research.

While several commercial sources of immortalized cell lines exist, two of the largest are the American Type Culture Collection (ATCC, www.atcc.org) and the Health Protection Agency Culture Collection (HPACC, also known as the European Collection of Cell Cultures, http://www.hpacultures.org.uk). These organizations house thousands of cell lines from a variety of origins, and are always a good starting point when acquiring cultured cell lines. Other cell types are available from biotechnology vendors, and some specialized cell lines are available only from particular institutes or individual investigators. Cells from ATCC and HPACC are typically classified by animal and tissue of origin, disease type (if applicable), morphology, growth properties, DNA profile, age/ethnicity of human source (if applicable), and so on. In most cases, the cell line is shipped as a cryopreserved aliquot (see Chapter 3 for detailed protocols on cryopreservation and thawing of cells), which must then be thawed and cultured. It is important to note that some cell cultures have stipulations for use. For example, use of the U-937 monocyte line [2] requires that the original paper is cited in all published work with that cell line. Listed below are some common lines used in cellular analysis. These cell lines do not represent an exhaustive list, but rather offer a few examples from several different cell types. The cells lines listed below are listed in no particular order, and only the parent (i.e., no mutants) cell line is discussed.

Chinese Hamster Ovary Cells. This double-X chromosome female cell line was derived from the Chinese Hamster (*Cricetulus griseus*). This cell line is proline-dependent, and has been transformed into several other clones (more than 30 mutants at ATCC alone) with different gene expression or transfection. The CHO cell line is epithelial in phenotype, and is used in applications involving microfluidic design and validation, and is also a popular host for transfection and genetic studies.

HeLa Cells. These cervical adenocarcinoma cells were isolated from an African American woman named Henrietta Lacks. Mrs. Lacks's cancer was so virulent that her cells, still used around the world today, have infiltrated many cell lines (see Chapter 3). Despite the vigorous growth of HeLa cells, and their propensity to cross-contaminate other cultured lines (e.g., the CCL-13 liver culture), they are useful both in tests of proliferation, and in viral studies. Since the cells replicate so rapidly, they can be used to generate large amounts of viruses, and can also be used to test anticancer activity of new compounds. Researchers using HeLa cells should, however, take precautions to isolate HeLa cell lines from other cell cultures to avoid cross-contamination. Like CHO cells, the HeLa cell line has been transfected into multiple clone types.

Jurkat Cells. Jurkat cells were derived from a 14-year-old male with acute T-cell leukemia. It is a suspended lymphoblast cell that expresses the CD3 antigen on its surface. Jurkat cells are relatively uniform in size for a cancerous cell line. Jurkat cells are T lymphocytes and, unlike some cancerous cell lines, readily undergo apoptosis using either the caspase-8 (mitochondria-mediated) or caspase-9 (receptor-mediated) pathways. Jurkat cells therefore are used as positive controls for apoptosis initiators and inhibitors. They are also useful for flow cytometry (either bench-top or microfluidic) as they are suspension cells.

U-937 Cells. U-937 monocytes are derived from a lymphoma patient. Originating from blood cells, they are suspended and therefore useful for microfluidic applications of cell separation, among other applications. Like Jurkat cells, they express the CD95 (Fas) antigen and are useful in apoptosis studies.

NIH-3T3 Cells. The NIH-3T3 cell line is an embryonic mouse fibroblast. It is used extensively in microfluidic testing as a model system for adherent cells [3], due in part to its genetic and morphological stability. The cell line is used as a feeder layer for other cultures such as keratinocytes.

Rat Basophilic Leukemia (RBL-1) Cells. The RBL-1 line is a lymphoblast and is used as a model for suspended-cell analysis [3,4]. It is a stable cell line that can be used for a variety of cytometry and separation applications. Amaxa (www.amaxa.com) has detailed protocols for green fluorescent protein (GFP) transfection of RBL-1 cells using their products.

HT-29 Human Colon Adenocarcinoma Cells. The colorectal adenocarcinoma HT-29 line can be used for a variety of cancer-cell studies, [5] as well as experiments using model adherent epithelial cells. This cell line is highly aneuploid (chromosome count rages from 68–72). The cells grow with a large variation in size and morphology, an indicator of their virulent nature. HT-29 cells are tumorigenic and should be handled with care to avoid cross-contamination.

HL-60 Cell Line. The HL-60 line is a suspended myeloblast that is used because it is a robust cell for microfluidic applications [6]. It produces tumor necrosis factor-α (TNF-α) upon stimulation [7].

Molt-3 and Molt-4 Cells. Molt-3 cells are leukemia T lymphoblasts that express CD4 (among other antigens), making them excellent models for CD4 + T lymphocyte separation or sorting models (Chapter 5). They can be used in development of CD4 + T cell isolation by microfluidic FACS or MACS, or by affinity separations [8] before blood samples are evaluated. The same leukemia patient produced the Molt-4 T lymphoblast cell line, which expresses CD4 as well as CD3 (among others). Molt-3 and Molt-4 cell lines have different numbers of chromosomes, and both can be used for a variety of separation, flow cytometry, and FACS applications (see Chapter 5).

Human T Cutaneous Lymphocyte (HuT 78) Cells. The HuT 78 cell line is another CD4 + cell line that can be used for cytometry and cell separation applications [9]. This cell line has a more variable morphology than other lymphoblast lines discussed so far. HuT 78 cells are a robust line, but exhibit clumping behavior in culture, requiring mechanical disruption prior to cytometry applications. They produce interleukin-2 (IL-2) and TNF-α and are tumorigenic.

Raji Lymphocyte Cells. Raji cells are B lymphocytes acquired from a patient with Burkitt's lymphoma. They express CD19 and are therefore a good model for B cell isolation, FACS, and other separation strategies.

PC-12 Cells. The PC-12 cell line is an excellent model for both differentiation and neuronal studies. Isolated from a rat adrenal gland, this cell type has a polygonal morphology and clumping behavior in culture. Once activated with nerve growth factor, this cell line differentiates into neuron-like cells that display neuronal activity.

CCRF-CEM Cells. These cells are human T lymphoblasts that express the CD3 and CD4 antigens, among others. They are a good model for T-cell separations based on affinity separations, FACS/MACS, or other methods (Chapter 5).

HL-1 Cardiomyocytes. The HL-1 myocyte line was established by Claycomb and co-workers in 1998 [10] and several additional clones have been established since. The cell line is capable of continually dividing while retaining contractile capabilities. It is an excellent model for muscle studies (particularly atrial muscle).

C166-GFP Cells. This cell line is a clone of the C166 murine yolk sac endothelial cell line. It possesses the characteristics of its parent line, but has been transfected to express GFP. This line is useful for imaging of cells in microfluidic devices for cell culture [11] due to its high expression of GFP (as mentioned in Chapter 4, imaging of the cells in phenol-red-free medium will improve the image signal-to-noise ratio).

CCD-1064Sk Skin Cells. These fibroblast cells were obtained from the foreskin of a normal, newborn male. The line undergoes senescence after approximately 54 passages. The CCD-1064Sk cell line can be useful for studies of nonimmortalized cells, aging and senescence research, and genetic analysis (the chromosome number is normal).

Detroit 532 Skin Cells. This cell line can be paired with the aforementioned CCD-1064Sk line. Both cell lines are derived from foreskin cells of newborn males. The Detroit 532 line also undergoes senescence (after approximately 30 passages). The Detroit line is from an infant with

Down syndrome (trisomy 21) and the cells have 47 chromosomes. It is therefore a good control for genetic analysis when paired with the CCD-1064 skin cell line.

INS-1 Cells. These rat-tumor beta cells produce insulin in culture and are immortal [12]. They are useful models for a variety of diabetes and insulin secretion studies, as they are capable of releasing insulin in response to glucose concentration and physiological glucose levels.

1.5   CHOOSING PRIMARY CELLS

Unlike cultured cells, where one often seeks a model to demonstrate a cellular analysis, primary cultures are chosen for the biological activity tied to the animal of origin. If a method is to be validated and optimized, it is best to choose a cultured model rather than obtain primary cells. Primary cells require a higher level of infrastructure (animal-care facilities) as well as regulation. However, what primary cells offer that cultured cells cannot is a higher degree of biological relevance, as the primary cells retain more of their phenotype of origin (this diminishes over time, see Section 1.9). It is therefore, at times, necessary to use primary cells, and this section will serve as a guide for obtaining and caring for these cells.

For all of the benefits of primary cells – such as correct phenotype and biological response – there are several obstacles to obtaining them. First, and most important, is the Institution Animal Care and Use Committee (IACUC, or similar acronym). The role of the IACUC at any institution is to approve animal protocols and ensure that animals used for research are treated humanely, safely, and without waste. Each university, hospital, research facility, and so on, will have a committee of this type, which is responsible for approving any animal protocols. In most cases, this committee must approve animal protocols before or during the application for extramural funding, so it is best to begin the acquisition of primary cells by finding and familiarizing oneself with this committee. The IACUC oversees all aspects of animal research, from behavioral to surgical. If the primary cells of interest are of human origin, then the appropriate Human Subjects Institutional Review Board (IRB) committee will be addressed instead.

The source of the animals/tissue will dictate largely how the IACUC process proceeds. For example, if one is obtaining tissue from an animal harvested for another purpose (e.g., an animal-sciences teaching lab), then use the tissue may fall under the other investigator's protocol, and one only needs to address the IACUC committee and document that the tissue was already harvested by an approved protocol. The benefit of collaborating in this manner is that the animal-care facilities are not housed in one's own laboratory, reducing costs. Some universities and institutes have dedicated campus-wide animal facilities to meet the needs of investigators across the facility. This approach has many advantages, such as centralized facilities, dedicated veterinary staff, and greater security. The latter point of security cannot be stressed enough. The issue of animal research continues to be debated strongly by its opponents and researchers, and acts of vandalism on animal facilities are a possibility, albeit a rare one.

The animal type will determine, in part, the rigorousness of the IACUC process. Invertebrate animals are rarely regulated by IACUCs. Mice and rats are the most common vertebrate research animals, in part due to their lifespan (several years) and similar biology to other mammals. Mice and rats have extensive antibody libraries, as well as many mutant strains for research. Using larger animals complicates both the IACUC and animal-care process (and increases costs). The location of the animals is also important. Housing the animals in one's own lab requires more regulatory control than centralized animal facilities. Regardless of the animal type and location, the following concerns (and others) must be addressed. These sections are based on the National Institutes of Health Vertebrate Animals Assurances for R01 Grant Applications, but they provide a framework for many IACUC procedures.

Animal Description. Are the animals bred for research or commercial use? Will both genders be used, and over what age range? Will mutants or animals raised under stress or special diet be used? These are just some of the questions one must address when describing the animals of interest, as they indicate to the IACUC what pool of animals will be used. For example, as one moves along the evolutionary chain, the degree of difficulty in obtaining animals increases. Fish will raise fewer questions from the IACUC than small primates. At the same time, unnecessarily including infant animals may also cause delays or problems with the approval process.

Justification of Animal Use. Why must tissue be harvested from these animals? How many will be used? One must make a clear, compelling case for why cultured cell models will not answer the questions of interest, or why the cells are crucial for an analytical technique to succeed. At the same time, an incorrect estimation of the number of animals needed may also hinder IACUC approval. Since biological variation is unavoidable, the differences between primary cell lots may be significant, necessitating larger sample sizes (and more animals) to generate statistically meaningful data.

Veterinary Care. In many cases, harvesting the animal for tissue or cells is not a survival procedure. Up until and during the moment of harvest (a term used increasingly as a replacement for "sacrifice") each animal must be cared for and its health documented. Again, having the animals in one's own laboratory requires regular site visits by the facility veterinarian. The care of the animals must be explicitly described to assure the IACUC of humane animal care for the entire project duration. As can be expected, the more complex the animal, the greater the care costs.

Procedures for Reducing Pain and Distress. As mentioned above, the isolation of most primary cells results in harvesting the animal. The exact protocols during care and during the harvesting process must be documented and followed to ensure humane harvesting of the animals. Methods of restraint, harvest, and handling of subsequent tissues and organs must be documented in detail. One must also assure the IACUC that these methods are consistent with the practices of the veterinary medical association of the country in which the work will be performed. From a practical standpoint, it is easier on the investigator to use a centralized facility or find a collaborator who already possesses the animal infrastructure. However, for smaller animals, such as rats and mice, it may be more desirable to house the animals in the lab in which the research will be conducted.

Human primary cells introduce another level of complexity and regulation. In general, most biologist and chemists will have to collaborate with medical personnel to obtain samples via surgery, and so on. Human cell research also requires that the gender, age, ethnicity, and health of the donors are all tracked in order to generate meaningful results about the population at large. Obtaining human cells also has additional ethical issues, especially if one intends to transfect these cells. Given the large number of animal models for most human diseases, the need to obtain human primary cells is rare, with the exception of human cells that are relatively easy to acquire (Section 1.6).

One of the most interesting sources of animal cells is the zebrafish (*Danio rerio*). The zebrafish embryo can be imaged intact, or the cells of the developing fish can be removed and purified. Since zebrafish embryos are relatively easy to obtain, they offer another avenue of primary cells.

1.6   EASILY OBTAINABLE PRIMARY CELLS

Of all of the primary cells available, those in suspension are the easiest to obtain. Whether animal or human in origin, these cell types can be readily harvested without harming the donor, although IACUC or Human Subjects IRB approval will still be required. Cells found in blood and in ejaculate are both readily obtainable in large numbers due to their high concentration in their carrier fluids. Since no tissue is excised, and the donor will survive the donation process, these cell types can be obtained from the same donor repeatedly if necessary, reducing lot variability. In addition, it is possible to follow dose–response experiments in the same donor over time, which is not possible if the animal does not survive the cell isolation process.

The benefits of using blood and sperm cells – where applicable – include ease of handling, the relative ease of obtaining samples, and the health impact of both cell types. The far-reaching importance of both blood and gamete research makes these cell types not only convenient to use (relative to adherent cells from tissue), but also quite relevant. For example, if a new, chip-based cell sorter is being developed, the ability to sort primary blood cells will not go unnoticed. There is continuing need to analyze blood from an immunology standpoint and for medical diagnosis, as well as for genetic testing. Sperm cells, like blood, have importance both in genetics and fertility research. Analysis of semen is difficult in part because of the viscosity and heterogeneity of the fluid, and due to the presence of many proteins in high concentration. Blood also presents a unique challenge in cell analysis. Relevant cell concentrations range from 10^9 per ml (erythrocytes) to 1 per ml. Searching for changes in leukocyte populations, or detecting circulating tumor cells (Chapter 5) requires discriminating between many cell types and their "normal" concentrations. In addition to the large span in cell concentrations, the number of cell types found in blood complicates many analyses. Also the presence of the diverse and complex plasma proteome can create problems for many analytical techniques. Therefore an analytical system that can handle whole (preferred) or lysed blood is one that is robust enough for most other cellular analyses.

Like any primary tissue, one must treat blood and semen from donors with the strictest of blood-borne-pathogen protocols (Section 1.13). The safest way to handle primary (and cultured) cells is to assume that they could contain a hazardous pathogen. This mindset should be constantly stressed in the laboratory, as familiarity with procedures is usually the breeding ground of cavalier behavior (and therefore accidents). Blood and other bodily fluids can be fixed to reduce (but not remove) the threat of blood-borne pathogens, although the primary cells are no longer viable at that point.

Obtaining blood and semen from animals requires a suitable IACUC procedure. Obtaining the same fluids from humans is more complicated, as any discovery about the sample (e.g., a state of disease) could be reflected back to the donor. Unless the testing for said disease is disclosed and the donor consents to the test, the blood samples must be masked from the donor identity. One method to circumvent this issue is to obtain blood from a third party that has a sample code that is not traceable (by the end user in this case) to the donor. Once the sample protocols have been set up as donor-blind and approved, then partnering with a local clinic or hospital is the most straightforward method of obtaining the samples.

Blood and sperm cells are largely terminally differentiated – although some leukocytes activate upon stimulation. Therefore healthy blood and sperm cells cannot be perpetuated in culture indefinitely, requiring that the samples be tested in a timely manner. This element of timing is nearly ubiquitous in primary cell research; once the samples are obtained, the experiment cannot be delayed significantly before the cells die or change in culture. If the viability or function of the cells is unimportant – such as in cell separations or many flow cytometry analyses – then fixing the cells allows the sampling and experiment to be separated in time, and also reduces blood-borne pathogens and changes in the cell phenotype over time. If fixed blood is acceptable to demonstrate a new analytical technique (i.e., proof of concept, rather than elucidating new information about blood cells themselves) then blood controls can be used [13]. Blood-cell controls such as the Multi-Check Control set of standard blood samples from Becton-Dickinson and other suppliers are fixed blood containing an anticoagulant in vacutainers. The samples are not identified with donors in any way, and the blood samples come with phenotype reports, which can be used to verify new cytometric analyses. Another benefit of using blood controls is that the blood can be purchased with depleted cell types, which requires MACS sorting or other sample treatment to produce in the laboratory.

1.7   PRIMARY CELLS FROM TISSUES

Unlike their suspended counterparts, most adherent cells require some degree of surgical isolation and disruption of the tissue before cells can be obtained. At this point, IACUC or Human Subjects approvals should have been obtained and the lab should have been cleared for the appropriate biosafety level (Chapter 2) and animal care (if applicable). Human and animal tissue will be treated the same in this discussion, since once they are excised from the donor they only differ based on physiology. The essential process for obtaining cells from living tissues is outlined in Figure 1.2. After the animal is sacrificed, the tissue of interest is excised for further processing (Figure 1.3). In the case of human research, the excision coincides with a surgery or biopsy. Tissue excision can include the whole organ or a sufficient section to obtain cells of interest. Generally, smaller animals will require whole organ excision unless a particular tissue type is targeted (e.g., ventricular vs. atrial myocardium). Excision is typically followed by perfusion with a buffer. This initial perfusion stage removes blood and other fluids from the tissue and helps to instill sterility in the sample. If muscular tissue is used, the muscles can be arrested by perfusing Ca^{2+}-free buffer [14]. The tissue is then cut into smaller pieces (if needed) and incubated with a digestion agent such as collagenase, trypsin, or papain. Centrifugation and separation steps then follow. The separation depends in part on what cell types are isolated. If an external marker

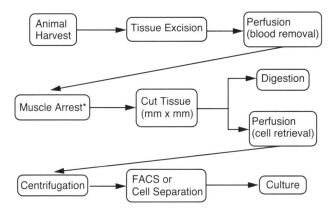

Figure 1.2   Basic steps in primary cell isolation. In the case of animals, the donor organism is typically harvested before tissue excision and removal of blood from the dissected tissue. In the case of muscle isolation, calcium-free buffer is perfused to arrest the muscle. Cut tissue is digested and purified by one of several isolation methods. (*Indicates optional step for muscle tissue)

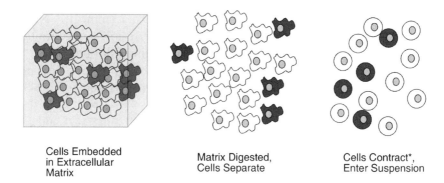

Cells Embedded
in Extracellular
Matrix

Matrix Digested,
Cells Separate

Cells Contract*,
Enter Suspension

Figure 1.3 Concept of tissue digestion. An excised piece of tissue, containing cells of interest (dark gray) and other, background cells embedded in an extracellular matrix (ECM). A digesting agent (e.g., trypsin) dissolves the ECM, releasing the cells. Cell contraction into suspended, near-spherical cells is reversed upon subsequent culture of the isolated cells. (*Not all cells contract into spherical shape)

specific to the target cells is known, then FACS, cell-affinity chromato-graphy, or MACS sorting methods are possible. It is also possible to use selective culture in some cases (see Section 1.8).

There are numerous protocols available for cell isolation from primary tissues. It would not be possible to devote a single chapter to every common or possible isolation protocol. Rather, four different isolations will be presented so that common themes and specific variations can be noted. The methods of sacrifice are not discussed as they vary by IACUC and institution. Methods to purify the final digests are discussed below. Investigators should consult the relevant literature for specific protocols for a given animal or tissue type.

1.7.1   Isolation of Cardiomyocytes from Mammalian Hearts

This procedure is typically performed on rat hearts [14–16], although it can also be used with larger animals. Cardiomyotyces represent a unique class of muscle cell and one that involves a structurally complex organ of origin. This particular procedure is modified from the work of Diez and Simm [16].

1. After harvest of the animal, the heart is removed. Adding heparin is recommended in the initial perfusion to reduce clotting. The excised heart is then attached to a Langendorff perfusion system and perfused with a buffer ($110\,\mathrm{mmol\,l^{-1}}$ NaCl, $2.6\,\mathrm{mmol\,l^{-1}}$ KCl,

$1.2\,mmol\,l^{-1}$ KH_2PO_4, $1.2\,mmol\,l^{-1}$ $MgSO_4$, $25\,mmol\,l^{-1}$ NaH-CO_3, $11\,mmol\,l^{-1}$ glucose) solution. This perfusion step continues until blood is no longer visible in the perfusing solution.

2. Ca^{2+}-free buffer is then perfused until the heart stops beating. Buffer with $0.25\,mg\,l^{-1}$ collogenase is then perfused for 20–30 minutes.

3. The heart is then removed from the Langedorff apparatus and cut into approximately $1\,mm \times 1\,mm$ pieces. At this point the atria and ventricles should be treated separately (if only one cell type is required, the other portions of the heart may be discarded appropriately).

4. The tissue pieces are then incubated for 30 minutes in buffer (from step 1, containing 400 mg albumin and 0.25 mg collogenase), then filtered ($250\,\mu m$ nylon) and centrifuged. The remaining pellet is then resuspended in buffer or medium containing penicillin and streptomycin.

5. The resultant cell mixture contains both cardiomyocytes and endothelial cells. Two methods can be used to purify the cardiomyocytes. Due to their elongated shape and large size, label-free FACS can be used to sort cardiomyocytes [16]. If FACS sorting is unavailable, selective culturing [17] can be used to deplete adherent fibroblasts and endothelial cells, while the supernatant (containing myocytes) is collected.

1.7.2   Isolation of Pancreatic Beta Cells from Rats

The rat pancreas has served as the standard model of human pancreatic function and is vital to the study of diabetes. Islets of Langerhans and the cells that reside within them are highly differentiated and not robust after isolation. While INS–1 rat tumor cells (Section 1.4, Table 1.2) can be used to conduct many glucose-dose–insulin-response studies, primary tissue will provide better physiological relevance, and the donor animals can be raised on special diets prior to isolation. The following procedure is adapted from Ni *et al.* [18] and Roper *et al.* [19], and outlines the general procedure for rat-islet isolation after the animal has been harvested.

1. After the animal is harvested, the pancreas is inflated by injecting collagenase XI ($0.5\,mg\,ml^{-1}$) into the duct of the pancreas.

2. The pancreas is then removed from the animal and placed in collogenase solution for 8–10 minutes to disrupt the extracellular matrix.

3. At this point two possible purification steps can be performed.
   a. Islets can be isolated using a nylon mesh filter to capture islets 100–200 µm in diameter.
   b. Exocrine and endocrine tissue can be separated using a Ficoll gradient (see Chapter 5 for a discussion of this technique). Islets are then selected by hand using a stereomicroscope and transferred for culture.
4. Antibiotic and antifungal agents are added to the medium to resist contamination.

1.7.3   Isolation of Skin Cells

Skin cells vary by function and differentiation. In addition to the heterogeneity of the tissue, the presence of foreign particulates, bacteria, and other contaminants makes purification and sterility more challenging than some cell types. These difficulties are offset by the ability to obtain tissue from donors in a nonlethal manner. Human skin cells are therefore relatively easier to obtain than other tissues. The ability to obtain cancerous/diseased tissue as well as healthy/control tissue from the same donor also makes primary skin cells attractive choices. The following protocol for human skin cells was adapted from Johnen *et al.* [20].

1. Prior to surgical removal, the skin should be clean and shaved to remove hair.
2. After the tissue has been removed from the donor, the dermal and epidermal layers are removed from each other by 16 h treatment with dispase solution.
3. Keratinocytes are then isolated from the epidermis using 0.05% Trypsin–EDTA (ethylenediaminetetraacetic acid) solution.
4. At the same time, a second sample from the same biopsy is used to isolate fibroblasts by treatment wit 0.4% collagenase.

1.7.4   Isolation of Neurons

For patch-clamping and other electrophysiological measurements, as well as some measurements of neurotransmitter secretion, it is desirable to obtain free neurons. Removal of neurons from their native tissue – free of glial cells – is therefore essential for many experiments. The following protocol, adapted from Kay and Wong [21] presents a straightforward method for neuron isolation. While the original protocol was for guinea pigs, it can be modified for other animal sources. A digestion chamber

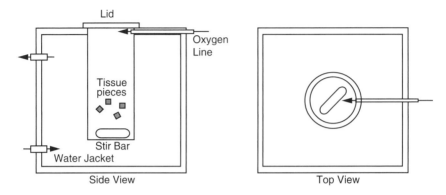

Figure 1.4 Digestion chamber for neuron isolation, adapted from [21]. The tissue pieces are incubated in a digestion buffer with minimal agitation using a stir bar. Cells isolated in this device can be maintained for up to 10 h after removal from the donor

(Figure 1.4) is used for incubation with enzymatic buffers. The chamber allows cells to be digested and then maintained for up to 10 h. Once isolated and removed from the chamber, the neurons only remain functional for 2 h.

1. Brain tissue – or the entire brain – is removed and placed in 5 °C PIPES saline with d-glucose. The tissue of interest is dissected with a scalpel and cut into approximately 0.5–1.0 mm pieces. As much of the white matter is removed as possible at this stage.
2. A small paintbrush is recommended to transfer slices, rather than tweezers, to avoid pressure on the brain tissue. Tissue pieces are placed into the digestion chamber's inner volume (Figure 1.4). The chamber contains PIPES saline and trypsin. The chamber is held at a constant temperature (32 °C) and oxygenated (100%).
   Note: The stir bar in the central chamber (Figure 1.4) should be operated slowly. The tissue should reside in suspension above the stir bar. Tissue slices should not appear mechanically damaged, or the stir rate should be decreased further.
3. After 1.5 h, the solution is removed (the tissue is retained in the chamber) and replaced with PIPES saline (see [21] for exact concentrations) and oxygenated. The solution is allowed to equilibrate to room temperature.
4. Neurons should be removed from the chamber when needed. Two pieces of tissue are removed and placed in Dulbecco's modified Eagle's medium (DMEM, with HEPES (4-(2-hydroxyethyl)-1-piperazineethanesulfonic acid) and D-glucose). Two Pasteur pipettes

(one of 0.5 mm diameter and the other 0.2 mm) are used to triturate the tissue every two seconds. The larger pipette is used until the tissue passes into it without difficulty; the second, smaller pipette is then used in the same manner until the slices pass into it easily.

5. The solution is placed in a Petri dish or similar and the neurons settle. After neurons have attached, the solution and suspended materials are removed and replaced with HEPES saline. The cells are then ready for analysis.

The protocols listed here are four examples of the hundreds of protocols available for primary cell isolation. The animal harvesting process (if applicable), dissection, and disruption of cells differs largely based on tissue and cell type. The up-front regulatory and approval work is necessary and important for minimizing harm to the donor, and to ensure that ethical and reproducible experiments are conducted. However, the end result is that physiologically relevant cells can be obtained in a straightforward manner.

1.8   PURIFYING PRIMARY CELLS

Chapter 5 discusses in-depth methods of cell separation. However, a brief discussion of these techniques is presented here since most cell isolation experiments require at least one separation step. The reason for this separation step is clear when one considers the isolation of cells from a donor tissue (Figure 1.3). The original tissue is a mixture of cell types and extracellular matrix, as well as connective tissue. After mechanical disruptions and enzymatic digestion, the remaining viable cells are in suspension and must be separated from each other. Some methods of cell selection are mechanical, others chemical (see Figure 1.5 for criteria that can be used to select a cell-separation method). Not all methods discussed in Chapter 5 can be used on primary cells that will be used for subsequent experiments. Some methods, while useful for some isolated cells, cannot be used for more mechanically sensitive cells. Therefore, a short overview of techniques can be used to select a method of cell isolation; for other cell-selection methods and in-depth discussions of the following techniques, the reader is referred to Chapter 5.

Selective Culture. In some cases, such as the isolation of cardiomyocytes from supporting endothelial cells, the difference in cell buoyancy or adherence can be used to isolate cells. If a suspended cell is perfused

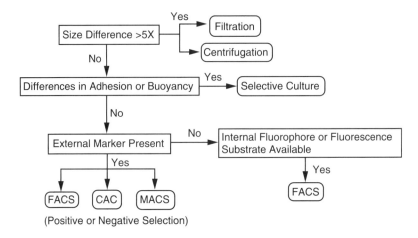

Figure 1.5 Decision process for selecting the correct method of purification for primary cells. Size, buoyancy, and external or internal markers can be used to isolate cells of interest. Filtration, centrifugation, and selective culture take advantage of differences in the physical properties of target and background primary cells. FACS, MACS, and CAC use internal or external markers to isolate the target cells. Multiple methods may be used in tandem for higher purity. FACS = fluorescence-activated cell sorting; MACS = magnetic-activated cell sorting; CAC = cell-affinity chromatography

out of a tissue comprised of adherent tissue, then the digest can be cultured on Petri dishes and the adherent cells allowed to settle and attach. After several hours, when adherent cells can be seen to spread and attach under microscopic investigation, the supernatant is removed and transferred to a fresh dish. The process is repeated several times so that the few remaining adherent cells can be removed. Selective culture can also be used in the reverse mode to remove suspended or slowly attaching cells from the target cell population. Regardless of which mode is chosen, this separation method is time intensive, and only works when a difference in cell settling or adhesion exists. However, it is a simple procedure, gentle on sensitive cells, and far less expensive than a fluorescence-activated cell sorter.

A variation of this technique is to use affinity capture to selectively remove cells [17]. Rather than rely only on gravitational settling to deplete unwanted cells, a ligand that is selective for the undesired cell type(s) is used to promote surface adhesion while the target cell remains in suspension. This approach requires some type of external surface marker that is not found on the target cell type. Negative depletion of this sort may be faster than settling and cell attachment, and may work when both target and nontarget cells both tend to settle and attach in similar time frames. The benefit of selective culture – with or without a capture ligand – is that the target cell is enriched without the use of labels (which can sometimes affect cell function in downstream experiments).

Filtration. As discussed in Chapter 5, a significant difference in size or shape of two or more cell types makes mechanical separation straightforward. Filtration is a useful step in cell purification to remove cell clusters and other large particulates. Filtration can also be used to separate target and nontarget cells; however, when one considers the large number of cell phenotypes in a given organism, few cells deviate significantly in size for adequate separation. Aside from cross-flow microfluidic methods (Chapter 7) to separate red and white blood cells, the size difference required for other mechanical filtration must be large. For example, the separation and isolation of islets – which are multi-cellular "units" – is possible using nylon mesh. Other cell types that may be isolated include myocytes, which are typically elongated relative to other cells, or other multi-cellular units such as ganglia, and so on. Red and white blood-cell separation is another example of filtration, although microfluidic devices are more successful than nylon mesh at those size scales. The benefit of filtration is that large volumes of cells can be processed, the separation can be repeated many times in a short timeframe, and filtration is not expensive. However, given the few cases where filtration is the best method, one cannot rely on filtration alone for all cell isolation needs.

Fluorescence-Activated Cell Sorting (FACS). Aside from the microfluidic FACS devices discussed in Chapters 5–7, the majority of FACS instruments are large and expensive, and will likely be housed in a central facility of some sort. Chapters 5 and 6 discuss the FACS mechanism in detail. Briefly, differences in light scatter and/or fluorescence are used to identify a target cell. Cells are then dispersed into droplets, which are deflected into collection or waste chambers. High-speed sorting instruments can sort an excess of 10 000 cells per second. FACS sorting is a high-speed method that can take advantage of multiple scatter and fluorescence parameters to achieve high purity and high yield. However, FACS methods are not always ideal, or even advisable in some experiments. The FACS sorting mechanism most instruments use is electrostatic deflection of droplets, which can damage sensitive cells. Even fluidic sorters can achieve higher shear forces for delicate cells. Another drawback of FACS methods is that the goal of the separation is to produce viable cells that can be cultured in a sterile environment. Therefore the cell must not only survive the sorting process, but also must survive in the collection vessel long enough to reach culture. The final concern for any FACS analysis is finding unique surface markers or scatter parameters to sort cells. While FACS separations can take advantage of intracellular antigens, it is not always feasible to access these and other internal analytes while maintaining cell viability.

An example of FACS purification using external markers only is the isolation of primary cardiomyocytes [16]. Rather than use a specific surface marker – which must then be removed before culture or analysis in some cases – the amount of autofluorescence, the time of flight through the laser beam, and the scatter can be used to discern target cells. The autofluorescence and scatter are both indicative of cell size, and the time of flight (duration of fluorescence pulse) indicates the length of the cell. Using these parameters, the rod-shaped, elongated cardiomyocytes can be sorted from endothelial cells from the same tissue digest.

Magnetic-Activated Cell Sorting and Cell-Affinity Chromatography Methods. Like FACS separations of primary cells, magnetic-activated cell sorting (MACS) and cell-affinity chromatography (CAC) both require a unique surface antigen to capture target cells. Both methods can also be used to deplete nontarget cells. As discussed in Chapter 5, MACS separations use magnetic microparticles or nanoparticles to isolate a cell type in a magnetic field, while unlabeled cells are removed to waste (Figure 5.2, Chapter 5). In CAC, capture ligands bound to a support capture the labeled cells while unlabeled cells flow to waste (Figure 5.8, Chapter 5). If a target molecule on the cell surface can be identified, then the target cell can be enriched – or nontarget cells depleted – using either method. In both cases, depletion of unwanted cells will leave the target cell free from any labels. The benefit of both CAC and MACS is that both can be adapted to microfluidics and both can be operated in continuous, flow-through methods for large volume throughput during cell enrichment. CAC can also be operated in a closed, sterile manner, reducing the risk of contamination (Chapter 5).

1.9 HOW LONG DO PRIMARY CELLS REMAIN PRIMARY?

The reason for using primary cells is to harness the physiological functions of the cells in question. Cultured (immortalized) cells that retain true physiological function are the exception rather than the standard; therefore, primary cells and animal harvest are currently the best approaches.

However, the timing of primary-cell acquisition, experiment duration, and numbers of experiments do not always match precisely. It is necessary, therefore, to maintain primary cells in culture for short durations. The time that primary cells, once purified, can be kept in culture depends on two factors. First, the lifespan of the cells outside of the body will be limited in some cases, especially for nonproliferating cells. Second, even proliferating primary cells will reach a stage at which point they either become senescent or they lose their *in vivo* phenotypes. An excellent example of the latter case is the loss of *in vivo* morphology – and in some cases contractile capability – in rat cardiomyocytes. Primary rat cardiomyocytes can be removed from the donor animal and cultured in the appropriate medium for an average of 9–10 days [22,23]. After this point, the cardiomyocytes lose their characteristic rod shape and resemble standard adherent cells, such as endothelial cells. The cells also lose their contractile behavior at this time. It is possible to stimulate contraction again, but after the 9–10 day culture period it would be better to procure fresh primary cells. Other cell types are less robust than heart muscle cells and will last shorter periods in culture. It is best, when establishing protocols for primary cell isolation and culture, that the initial primary cells be inspected daily for morphological changes and monitored for changes in physiology. For example, myocytes can be monitored for characteristic electrophysiology and contraction; islets can be monitored for insulin response to glucose challenges, and so on. In some cases, it is possible to stimulate proliferation of otherwise senescent cells [24]. Murine lymphoid cells [25] can be maintained in culture for several months to two years. In addition to acquiring new primary cells from another donor, another option is the use of cryopreservation (Chapter 3) to store multiple aliquots of primary cells from a given animal, thus minimizing any potential change to the cells from prolonged *in vitro* culture. Of course, with cryopreservation one must ascertain that the freezing and recovery process did not alter cell function.

1.10   OBTAINING PRIMARY CELLS FROM A COMMERCIAL SOURCE

Several companies – such as the Gibco branch of Invitrogen, StemCell Technologies, or Asterand – offer primary cells. These cells (see Table 1.3) are offered as cryopreserved or delivered in culture flasks. Using primary cells obtained from commercial sources does not typically require Human Subjects or IACUC approval, although each institution should be checked

Table 1.3 Some examples of commercially available cell lines

Tissue source	Organism	Vendor	Notes
Peripheral Mononuclear Blood	Human	StemCell Technologies	
CD34 + Stem Cells (Blood)	Human	StemCell Technologies	
Neurons	Rat	Gibco/Invitrogen	
Umbilical Vein Endothelium	Human	Gibco/Invitrogen	
Pulmonary Artery Endothelium	Human	Gibco/Invitrogen	
Mammary Epithelium	Human	Gibco/Invitrogen	
Epidermal Myelocytes	Human	Gibco/Invitrogen	Adult and neonatal cells available
Coronary Artery Smooth Muscle	Human	Gibco/Invitrogen	
Carcinoma-Associated Fibroblasts	Human	Asterand	Available from breast, colon, head, lung, and prostate
Chondrocytes	Human	Asterand	Available from normal donors and several different arthritis subtypes
Epithelial Cells	Human	Asterand	Head/neck, breast, lung, and prostate
Synovial fibroblasts	Human	Asterand	Normal and arthritic

for proper procedures. The number of primary cell lines is limited, although in some cases it may be easier to obtain commercially available human primary cells, especially those from donors with a particular disease state. For those interested in a readily available cell type, these commercial sources may provide a faster avenue to experimentation.

1.11 BACTERIA AND YEAST

While the majority of this book deals with mammalian cells, there are a large number of analytical techniques, particularly microfluidic ones (Chapter 7), that have been demonstrated with bacteria and yeast. The importance of both cell types is evident when one considers the encompassing concerns of contamination, food safety, defense, and the biotechnology industry. Many aspects of cell separation (Chapter 5), microscopy (Chapter 4), and flow cytometry (Chapter 6) apply to both mammalian and other cell types. An additional benefit of using bacteria and yeast is that they are readily transfected to produce fluorescent proteins and other biological products (many strains are already available producing several products of interest). Just as with mammalian cells, many vendors supply a variety of stains of bacteria and yeast cells. Unlike vertebrate animals (or even invertebrate ones), there are few, if any, regulations with respect to the health and welfare of prokaryotes. There are, however, strict regulations regarding the health and safety of the individuals who work with these cell types. Another issue that must be considered (see Chapter 3) is that the culture of prokaryotes in a lab that also cultures mammalian cells increases the risk of contamination to the latter.

1.12   PRACTICAL ASPECTS OF CELL CULTURE

Whether immortalized cells or primary cells are used in analyses, they must be maintained – even for short periods, in the case of some primary isolations – in culture. As discussed earlier, the culture of primary cells is limited by either the lifespan of the isolated cells or by the time between isolation and deviation of the cells from the original phenotype. This timeframe ranges from hours to weeks, depending on the cell type, species, and other factors. For immortalized cells, the culture can be continued indefinitely, although not without a new set of consequences. The majority of immortalized cells either come from cancerous tissues or have been modified to remain in the active cell cycle permanently. In both cases, the cells have deviated from their original state (presumably healthy cells and tissue). In adult donors, regardless of species, most cells have reached a steady state and are only replaced when cells are lost. Some cells are not replaced at all, and others enter a senescent state. Aging is attributed in part to the long-term loss of cells and tissue that are not replaced by mitosis. Consider, for example, the heart and brain, where neurons and cardiomyocytes, respectively, are lost over time and not replaced. Blood, however, is continually replenished and indeed does not show significant signs of aging throughout the lifespan of the individual or animal. Skin, on the other hand, exhibits self-repairing and the resilience of youth for a time, then loses the ability to replenish cells as rapidly (or at all).

Cancerous tissue, on the other hand, shows unlimited potential to replicate and grow. In humans this effect leads to tumor growth, but the ability for cancer cells to divide has led to new capabilities in science and medicine. The hybridoma, for example, has led to advances in monoclonal antibodies that have positively impacted the fields of flow cytometry (Chapter 6), fluorescence microscopy (Chapter 4), and cell separations

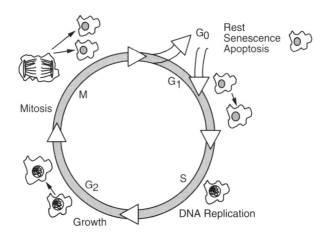

Figure 1.6   The mammalian cell cycle. Cells in the resting state (G_0) enter the growth phase (G_1) prior to DNA replication (S phase). The cell continues to grow (G_2) and finally divide by mitosis (M), producing two daughter cells. Immortal cells remain in the cycle (i.e., M → G_1) rather than enter the rest phase (G_0). Senescent and apoptotic cells exit the cell cycle and either cease to divide or undergo programmed cell death, respectively

(Chapter 5). Obtaining cells from cancerous tissue allows the cell type to replicate indefinitely as an immortalized cell line. However, cancer cells do not share the same phenotype as their healthy counterparts. The morphology, growth rates, antigen expression, as well as ploidy (chromosome number) can differ. Indeed, in the same culture over time, these and other factors can change. Therefore, a continuously passaged cell line should be checked periodically against frozen stock for changes in morphology, and so on.

The cell cycle for somatic cells is shown in Figure 1.6. Continuously refreshed cells remain in the cell cycle, while many somatic cells exit the cycle and become senescent. To produce a stable cell line from noncancerous cells or tissue, the cells must be isolated from a donor and then immortalized. Senescence – and aging – are governed by many factors, but one that has been positively linked to senescence is the shortening of the telomeres. As the cell replicates as part of the cell cycle, the telomeres lose base pairs, shortening with each replication. Cells that express telomerase, which rebuilds the telomeres, replicate indefinitely. Stem cells, germ cells, and many cancer cells express telomerase. Senescence is in part a defense mechanism for cells to avoid entering a cancerous state, as continued damage to DNA during the cell cycle would result in gene alteration.

From a cell-culture standpoint, immortalizing a cell line is an attractive option, as the cells of interest from a particular donor could be cloned indefinitely. The variability between organisms would then be removed, and a sufficient number of cells could be grown for a particular analysis. However, the immortalization process should ideally not alter the primary cells in any way. The process of culturing the primary cells outside of the donor tissue already introduces changes in morphology and cell function in some cells over time. Therefore, any additional changes should be avoided. For example, naturally occurring cancer cells – albeit immortal – rarely mimic the behavior of their tissue of origin. For laboratory-made immortal cells, the goal is to produce a stable, replicating cell line that retains the key feature of interest. There are several methods to immortalize cells, including viral transformation and recombinant transfection to produce telomerase.

In viral transformation, Epstein–Barr, Human papilloma virus (HPV), and other viruses are used to immortalize a cell and extend the replicative capabilities of the cell line. This process, while producing immortal and nearly immortal (a large number of passages before senescence) cells, does alter the cell function to an extent. In some cases, the alteration to ploidy and morphology is minimal; in other cases, there is a large difference between the original and transformed cell line. An alternative approach is to use human telomerase reverse transcriptase (hTERT, or TERT) technology. In this approach, the cell line is transfected to express telomerase (see Section 1.14 for transfection methods, and Chapter 9 for an hTERT protocol). The TERT process allows many somatic cells to become immortal, but alters the remaining biochemical machinery – cell function, ploidy, and so on – to a lesser degree than some viral transformations. Somatic cells contain the gene to express telomerase, but do not express the protein. Adding a plasmid to produce telomerase allows the cells to manufacture the protein regardless of the regulatory signals that suppress expression of the native telomerase gene. TERT-immortalized cells are purported to exhibit more native morphology and ploidy characteristics than viral-transformed immortal cells. However, the cells may still lose their *in vivo* phenotypic characteristics from the culture process alone (i.e., removal from their native tissue and deprivation of regulatory factors).

1.13   SAFETY ASPECTS OF PRIMARY AND TRANSFORMED CELL LINES

Regardless of the type of tissue, the source of origin, or the steps that subsequently follow in an analysis, the danger of exposure to hazardous pathogens is significant when working with cells. Just as proper chemical experimentation requires good hygiene and safety practices, so does cell analysis require appropriate safety procedures. The analytical cell laboratory (see Chapter 2) requires proper maintenance and protocols to ensure personnel safety. Personal protective equipment (also discussed in Chapter 2) must be used at all times and proper isolation, containment, and sterilization practices must be adhered to with strict regulation. As in any field in which a hazard exists, it is often the seasoned researcher who is injured. In the case of cell analysis, the risk of injury is so great – resulting in death in the worst cases – that safety cannot be overly stressed.

The Unites States Center for Disease Control and Prevention (CDC) has published guidelines for biosafety entitled *Biosafety in Microbiological and Biomedical Laboratories* (BMML). The BMML guidelines are available in print and online [26]. The guide covers biosafety levels, types of infectious agents, methods of containment and risk mitigation, and a host of other topics. This book should be placed in every cell-analysis lab and referenced when needed. It contains information pertaining to US regulations, but also gives general best practices for laboratories. Researchers in other countries should consult their own regulatory agencies for the exact rules regulations of their government. Nevertheless, the BMML guide is a useful tool for labs around the world.

The BMML guide and other regulatory documents discuss the need for *universal precautions* when handling biological materials such as cells and tissues. The concept of universal precautions is that one must assume that a biological material *does contain* infectious agents, even if it does not. The researcher is therefore always on guard to the potential threat of infection. It is the moment of distraction or relaxation of safety protocols when an injury occurs, not only to the individual, but also to others in the laboratory as well. Safety is discussed in greater detail in the following chapter, but should be considered at all times.

1.14   TRANSFECTION OF PRIMARY AND TRANSFORMED CELL LINES

Whether cells are cultured (immortal) or harvested from a donor, there is in some cases a need to add a reporter or otherwise transfect the cell. Transfection can be used to add a fluorescent protein reporter for protein expression studies, or to immortalize a cell through TERT transfection (see Chapter 9 for protocols for both). The introduction of exogenous DNA into a cell requires construction of the DNA vehicle itself as well as a cell introduction method. It is beyond the scope of this text to discuss plasmid construction, but there are several commercial sources of plasmids for a variety of reporters. There are several methods for inserting plasmid DNA constructs into cells, including viral vectors, lipid-based delivery, and electroporation (Chapter 9).

Viral transformation is a well-established technique that has been used to transfect many cell lines with reporters and other gene alterations. The main drawback of using retroviral vectors to introduce DNA constructs is the risk that the presence of the virus may affect cell function and may raise the biosafety level of the cell line in question (this is not always the case). Viral transformations work in a variety of cell types, both mammalian and bacterial. As an alternative to viral transformation, the physical methods of lipid delivery and electroporation do not require introduction of viruses or other organisms to achieve transfection. There are several lipid-based delivery options available, and most work by similar principles. The DNA construct of interest is encapsulated into a lipid bilayer vesicle. The vesicles are then incubated with the cell sample, where membrane fusion occurs. This method minimally stresses the cells – although in any transformation experiment, viability and cell function must be optimized – and other reagents can also be introduced into the cell. For example, lipid-based delivery can be used to introduce large proteins or quantum dots – or other nanoparticles – into cells. The Lipofectamine reagent is an example of commercial systems for lipid fusion for transfection, and is used routinely for cell transformation. It is also possible to introduce larger molecules into cells using a short peptide sequence of repeating arginine units [27].

Electroporation of cells requires no additional chemical or biological reagents other than the DNA construct in most cases, and can also be used to deliver larger molecules or particles as well. Electroporation uses a pulsed electric field to temporilily disrupt the cell membrane so that reagents of interest can enter the cells. Electroporation can occur on the single-cell level or can be accomplished for an entire culture of cells simultaneously. An electroporation protocol is discussed in Chapter 9, but the basic principles are discussed here. The Nucleofector system from Amaxa is an example of a commercial system for electroporation. Electroporation systems can also be purchased from other vendors (e.g., Harvard Apparatus). A home-built system can be made, although many electroporation systems offer greater control and some are approaching a purchase price that rivals many home-built systems. In electroporation, the field strength, current, pulse frequency, pulse width, and duty cycle all affect transfection efficiency and cell function. The same parameter set used for a particular cell line may not be the optimal set for a different cell type. If the cells are to be transfected with a fluorescent protein, the expression of the protein and subsequent fluorescence serves as a good marker of transfection efficiency and cell function. However, direct measurement of fluorescent protein reporters only yields the efficiency of protein production, as the gene delivery could be 100% efficient, but the expression of the protein could be lower. It is possible to introduce cell-impermeant fluorescent dyes or fluorescent proteins (already expressed and modified, and different from the encoded protein) along with the DNA to directly measure delivery efficiency (see Chapter 9).

1.15   CONCLUSION

In some instances, the cell type/function drives the analysis. In others, a new analytical method requires a cell type (or several types) for validation and optimization. The acquisition of cells can therefore serve as the driving force or an enabling step in analysis. In the case of primary cells, the application typically dictates the species, tissue source, and cell type. Primary cell isolation requires not only careful lab procedures, but also properly documented (and followed) procedures. The regulation of primary cells is important not only for experimental consistency, but also for the protection of the health and safety of the donor. In some cases, immortalization of the primary cell or tranfection may be needed to change the functionality of the cells. When cultured (immortalized) cells can be used, the variability is reduced, as are the regulatory procedures. Cultured cells often do not mimic their source-tissue counterparts, and therefore usually serve as models to validate or calibrate a method. Nevertheless, they are an important aspect of cell analysis. This chapter has covered many cell types, as well as basic protocols for primary cells. Subsequent chapters discuss use of these and other cells in a variety of analyses, as well as the practical aspects of maintaining cells in culture and ensuring sterile conditions that are free of contamination.

REFERENCES

1. Zhang, B., Adams, K.L., Luber, S.J. *et al.* (2008) Spatially and temporally resovled single-cell exocytosis utilizing individually addressable carbon microelectrode arrays. *Analytical Chemistry*, 80, 1394–1400.
2. Sundstrom, C. and Nilsson, K. (1976) Establishment and characterization of a human histiocytic lymphoma cell line (U-937). *International Journal of Cancer*, 17, 565–577.

3. Rettig, J.R. and Folch, A. (2005) Large-scale single-cell trapping and imaging using microwell arrays. *Analytical Chemistry*, **77**, 5628–5634.

4. Sims, C.E., Meredith, G.D., Krasieva, T.B. *et al.* (1998) Laser-micropipet combination for single-cell analysis. *Analytical Chemistry*, **70**, 4570–4577.

5. Hu, S., Zhang, L., Krylov, S., and Dovichi, N.J. (2003) Cell cycle-dependent protein fingerprint from a single cancer cell: image cytometry coupled with single-cell capillary sieving electrophoresis. *Analytical Chemistry*, **75**, 3495–3501.

6. Chang, W.C., Lee, L.P., and Liepmann, D. (2005) Biomimetic technique for adhesion-based collection and separation of cells in a microfluidic channel. *Lab on a Chip*, **5**, 64–73.

7. Aggarwal, B.B., Kohr, W.J., Hass, P.E. *et al.* (1985) Human tumor necrosis factor. Production, purification, and characterization. *Journal Biological Chemistry*, **260**, 2345–2354.

8. Revzin, A., Sekine, K., Sin, A. *et al.* (2005) development of a microfrabricated cytometry platform for characterization and sorting of individual leukocytes. *Lab on a Chip*, **5**, 30–37.

9. Wang, K., Cometti, B., and Pappas, D. (2007) Isolation and counting of multiple cell types using an affinity separation device. *Analytica Chimica Acta*, **601**, 1–9.

10. Claycomb, W.C., Lanson, N.A., Stallworth, B.S. *et al.* (1998) HL-1 cells: a cardiac muscle cell line that contracts and retains phenotypic characteristics of the adult cardiomyocyte. *Proceedings of the National Academy of Sciences USA*, **95**, 2979–2984.

11. Liu, K., Dang, D., Harrington, T. *et al.* (2008) Cell culture chip with low-shear mass transport. *Langmuir*, **24**, 5955–5960.

12. Eckholm, R., Ericson, L.E., and Lundquist, I. (1971) Monoamines in the pancreatic islets of the mouse. *Diabetologia*, **7**, 339–348.

13. Wang, K., Marshall, M.K., Garza, G., and Pappas, D. (2008) Open-tubular capillary cell affinity chromatography: single and tandem blood cell separation. *Analytical Chemistry*, **80**, 2118–2124.

14. Lee, S., Lee, H.G., and Kang, S.H. (2009) Real-time observations of intracellular Mg2 + signaling and waves in a single living ventricular myocyte cell. *Analytical Chemistry*, **81**, 538–542.

15. Jacobson, S.L. and Piper, H.M. (1986) Cell cultures of adult cardiomyocytes as models of the myocardium. *Journal of Molecular and Cellular Cardiology*, **18**, 661–678.

16. Diez, C. and Simm, A. (1998) Gene expression in rod shaped cardiac myocytes, sorted by flow cytometry. *Cardiovascular Research*, **40**, 530–537.

17. Davidson, M.M., Nesti, C., Palenzuela, L. *et al.* (2005) Novel cell lines derived from adult human ventricular cardiomyocytes. *Journal of Molecular and Cellular Cardiology*, **39**, 133–147.

18. Ni, Q., Reid, K.R., Burant, C.F., and Kennedy, R.T. (2008) Capillary LC-MS for high sensitivity metabolomic analysis of single islets of langerhans. *Analytical Chemistry*, **80**, 3539–3546.

19. Roper, M.G., Shackman, J.G., Dahlgren, G.M., and Kennedy, R.T. (2003) Micro-fluidic chip for continuous monitoring of hormone secretion from live cells using an electrophoresis-based immunoassay. *Analytical Chemistry*, **75**, 4711–4717.

20. Johnen, C., Hartmann, B., Steffen, I. *et al.* (2006) Skin cell isolation and expansion for cell transplantation is limited in patients using tobacco, alcohol, or a exhibiting diabetes mellitus. *Burns*, **32**, 194–200.

21. Kay, A.R. and Wong, R.K.S. (1986) Isolation of neurons suitable for patch-clamping from adult mammalian central nervous systems. *Journal of Neuroscience Methods*, **16**, 227–238.
22. Piper, H.M., Jacobson, S.L., and Schwartz, P. (1988) Determinants of cardiomyocyte development in long-term primary culture. *Journal of Molecular and Cellular Cardiology*, **20**, 825–835.
23. Volz, A., Piper, H.M., Siegmund, B., and Schwartz, P. (1991) Longevity of adult ventricular rat heart muscle cells in serum-free primary culture. *Journal of Molecular and Cellular Cardiology*, **23**, 161–173.
24. Halvorsen, T.L., Beattie, G.M., Lopez, A.D. *et al.* (2000) Accelerated telomere shortening and senescence in human pancreatic islet cells stimulated to divide *in vitro*. *Journal of Endocrinology*, **166**, 103–109.
25. Burker, T.R., Moore, G.E., and Stobo, J.D. (1975) Identification of lymphoid cells in cultures of murine leukocytes and thymus. *Cancer Research*, **35**, 673–678.
26. Center for Disease Control BMML Guide, http://www.cdc.gov/OD/ohs/biosfty/bmbl4/bmbl4toc.htm.
27. Dietz, G.P.H. and Bahr, M. (2004) Delivery of bioactive molecules into the cell: the trojan horse approach. *Molecular and Cellular Neuroscience*, **27**, 85–131.

2

The Cell-Culture Laboratory (Tools of the Trade)

2.1 INTRODUCTION

The use of cells in analytical chemistry, engineering, and biology requires a dedicated space for cell culture and maintenance. The proper handling of cells and tissues requires a level of diligence and constant education, to mitigate health and safety risks. Cell culture requires a system of mutual separation of sample and scientist to avoid contamination of either. Each time a culture flask and dish is opened is, in essence, an opportunity for a single bacterium or fungal cell to ruin an experiment. Likewise, every time cell cultures or tissues are handled, there is a risk to the scientist. It is therefore best to understand the protective countermeasures required to handle cells properly. As mentioned in Chapter 1, universal precautions assume that all cell cultures and related materials *may* contain hazardous pathogens. This assumption maintains a more vigilant attitude, and reduces the risk of accidental exposure to a real pathogen. Also, the possibility that cultures can be cross-contaminated requires additional – albeit similar – precautions. In short, careful procedures will result in productive research in a safe environment for cells and individuals. For those new to cells and cell culture, this chapter will not only serve as an introduction to the tools required for a cell lab, but will also detail some of the practical aspects to setting up a culture facility. For those with cell-culture experience, the discussion of analytical equipment should prove

Practical Cell Analysis   Dimitri Pappas
© 2010 John Wiley & Sons, Ltd

useful. Sterile culture techniques are also discussed. Discussion of medium, antibiotics, and so on, is found in Chapter 3.

2.2   ISSUES CONCERNING A CELL LABORATORY

Setting up a laboratory (or space within an existing lab) for cell culture is not a daunting task, but requires some planning and strict adherence to regulations. Most universities, institutes, and hospitals have a safety committee (some committees specialize in biosafety) that is in place in part to help an investigator establish a cell lab. While the government guidelines typically set the standard for safety rules, the investigator's institution may have additional guidelines to follow. The safety committee is therefore indispensable in the planning and setting up of a cell lab, as well as in the subsequent (and often frequent) safety inspections. The main issues when setting up and maintaining a culture lab are safety, sterility, and contamination. All three of these issues are linked by the common safe practices and proper use of equipment, and all three require that individuals working in the lab are properly educated.

2.2.1   Safety

Most institutions will require either a classroom or online training session to safely handle cells, tissues, or bodily fluids. Blood-borne-pathogen training consists of the types of potential pathogens (see Table 2.1), as well as methods to reduce the risk of exposure. In most cases, a physical barrier between the sample and individual suffices. However, in a lab with multiple users, the entire area must be treated as a potential risk, as telephones, computers, and so on, that should not be handled after touching culture materials, may have been contaminated. Safety training is typically repeated annually, although institutions differ in retraining frequency.

Both safety training and protocol training must take place before an individual is ready to work in a cell lab. Safety training concerns the mitigation of blood-borne pathogen risk to lab members. Protocol training, typically performed by more experienced members of the laboratory, ensures proper results and reduces the risk of contamination of samples. Both types of training should be repeated often, as familiarity and carelessness are closely linked. Proper cell-culture handling is discussed later in this chapter, and while specific methods of accident clean-up and safe disposal can be discussed, the variation in institutional requirements

Table 2.1 Blood-borne pathogens and exposure risks (nonchemical) in bodily fluids, tissues, and cell cultures

Pathogen	Associated risk(s)
Human Immunodeficiency Virus (HIV)	Acquired immune deficiency syndrome (AIDS), death.
Hepatitis B Virus (HBV)	Chronic liver disease, liver cancer
Hepatitis C Virus (HCV)	Chronic liver disease, liver cancer
Prions[a]	Prion diseases such as Creutzfeldt–Jakob Disease (neurodegenerative)
Ebola/Marbug (BSL 3 and higher)	Viral Hemorrhagic Fever

[a] Although not a living pathogen, the exposure risk when handling brain and spinal tissue should be considered.

is too great to encompass every safety rule in this book. Instead, some common biohazard accidents and mistakes are discussed, listed below.

1. **Exposure to open flasks/bottles, aerosol formation.** Every time a flask, bottle, or culture disk is opened, there is a risk of pathogen exposure. The same risk is introduced every time fluid is manipulated (pipetting, dispensing, pouring). For these reasons (and more) all cell handling should be performed in a laminar flow biosafety cabinet, also called a biosafety hood or laminar hood. These biosafety cabinets are discussed in Section 2.3. Even if the cell sample is about to be loaded into an unsterile instrument and sterility is no longer required, handling all cell-related fluid in a biosafety hood reduces risk of exposure. Safe handling of liquids using sterile pipetting techniques also minimizes aerosol formation, spillage, and so on.

2. **Needle Sticks.** Needle sticks are one of the largest sources of infection in both clinical and research labs. In the case of the latter, it is best to avoid their use whenever possible. However, there are several common rules for safe needle usage when it is required. A needle should remain capped until use. Over the years, there has been debate on what to do with a needle after the cap has been removed. Recapping the needle is generally regarded as prohibited, and with good reason. Recapping is one of the riskiest maneuvers when handling a needle. Other options have included a device to snip off the needle tip, although these are not allowed in some institutions as they generate an aerosol risk. It is best to immediately dispose of the exposed needle in a sharps container, and to move from sample (or patient), to analysis, to disposal, in one uninterrupted sequence. Excessive delay, or setting the needle down on a

surface, invites accidents. Blunt canulas, self-sheathing needles, and puncture-proof gloves (see personnal protective equipment for descriptions on nonstick technology) are all valid methods to reduce finger sticks in the laboratory.

3. **Centrifuge ruptures.** A less common, although no less dangerous biosafety accident is the failure of a centrifuge sample tube during centrifugation. Failure can occur from a poorly placed or damaged tube cap, or from a fault (i.e., crack) in the tube itself. An unbalanced rotor can also damage sample tubes from excess shock. Regardless of the type of failure, centrifuge rupture is particularly hazardous as the sample is not only exposed to open air, but often dispersed throughout the centrifuge chamber. Open tubes without caps are also prone to sample dispersion even if the tube has not failed, and should not be used with cell samples. If a centrifuge tube fails and sample is released into the system, the lid should be closed carefully and the system left alone for several minutes to let any aerosols settle. All sample tubes should be removed and discarded, even if they show no evidence of failure, as they may be contaminated by the ruptured sample. The centrifuge rotor should be removed carefully and its surfaces wiped with bleach. If an autoclave is available, the rotor should be sterilized (note: not all rotor materials are autoclavable, consult the manufacturer). If no autoclave is present, 10% bleach can be sprayed into the autoclave's tube holders to sterilize (the bleach is then poured into a container and disposed of according to the institution's chemical-hygiene rules). The centrifuge chamber should be cleaned with 10% bleach, with any cleaning agents (e.g., paper towels) disposed of in biohazardous waste. As a last precaution, sample tubes containing water should be run after complete sterilization to ensure the rotor is not faulty. Older centrifuge systems that do not have completely closed lids should not be used for blood or cell samples, as they may eject the sample beyond the centrifuge chamber upon failure.

4. **Dropped samples, tubes, and so on.** A dropped sample tube posses a splatter and aerosol hazard, and also results in loss of what may have been minutes to hours of sample processing. If a sample is dropped in an uncapped tube (such as a sample tube used in many flow cytometers), or if the cap or tube breaks, the lab should be cleared of personnel immediately to allow aerosols to settle. If the sample is dropped in a laminar biosafety cabinet, and none of the sample liquid splashed outside of the hood, then the laboratory does not need to be vacated. After a short (5–10 minute) time to allow

aerosols to settle, the spilled area should be cleaned with 10% bleach and paper towels (to remove the liquid and solid debris), followed by a second cleaning with 10% bleach (again, all paper towel waste should be disposed of as biohazardous material).

These four examples highlight some of the most common mistakes and accidents in a cell lab; however, they are not all-encompassing and there are many opportunities for biohazard exposure. If any lab personnel are exposed to cell cultures, blood, and so on, there are a set of safety protocols for every institution (and some common themes presented herein). It is advisable to learn of the exposure protocols *before* an accident occurs, including contact names/numbers and where individuals should go to report an injury of this type. The following example procedure highlights some of the common steps taken when reporting an accident in a cell lab; *again, one must follow the institutional guidelines when an exposure has occurred.*

1. Contact the lab supervisor *immediately*. If the principle investigator or lab director is not present, he or she must be notified as soon as the accident occurs. Even if the injury is not life-threatening (i.e., emergency services are not needed), the individual involved in the accident should report to the emergency room or designated clinic/facility for biohazard exposure. At this point, someone else in the lab should compile a list of all personnel present, the time of the exposure, and what was in the sample. It is important to list not only the cell/tissue type (species, etc.), but what other reagents were in the sample, if any. This information should be forwarded to the facility caring for the injured individual.
2. If blood-borne pathogens are expected to be present in the sample, the medical personnel may take a blood sample from the individual and test it immediately for any existing blood-borne pathogens as a baseline. If the contaminating sample can be recovered safely, that may be requested as well, for additional testing. If the sample came from a particular individual, that information should be documented and may be requested as well. If there is risk of chemical exposure from reagents in the sample, then there may be additional procedures required for the exposed individual(s).
3. The exposed individual will likely have to return several weeks or months later for additional testing to determine if infection has occurred. This testing may not be necessary if the original sample is preserved and safely brought to the health officials at the time of the

incident and original testing. For example, a dropped blood vial, which may have been exposed to a lab member, can be tested for the necessary blood-borne pathogens (Table 2.1). If no blood-borne pathogens are detected, then the exposed individual *may* not need additional monitoring (note: additional testing is at the discretion of the healthcare provider and any decisions or recommendations should be followed).

2.2.2   Sterility

Biohazard safety can be thought of, in a way, as a set of countermeasures to keep lab personnel from being infected by a specimen or sample. Sterility can therefore be regarded as steps taken to keep lab personnel and the surrounding environment from infecting the sample. Given the rich growth environment cell-culture medium provides, a single bacterium can ruin a cell culture or tissue sample. Indeed, when one considers the amount of time and funds required to maintain a culture, a persistent sterility issue can cause considerable damage. Contamination of cells by other cell lines is discussed later in this section. Here, the focus is placed on contamination by bacteria, fungi, and mycoplasma. Antibiotic and antifungal agents are discussed in Chapter 3; physical methods of sterility containment are discussed in this section.

Bacterial and fungal contamination is marked by depletion of medium nutrients and rapid proliferation of the invading agent relative to cell growth. In addition, it is next to impossible to conduct an experiment that yields any valid data in the presence of contamination. Bacterial and fungal contaminants can be mitigated through careful procedures, the use of antibiotic/antifungal agents (sparingly), and by proper sterilization of components. Mycoplasma contamination is more difficult to manage, as most reagents designed to eliminate bacteria rely on interaction with the cell wall, which is absent in mycoplasma. Bacterial contamination, in its later stages, is the easiest to detect. A bacteria-contaminated culture can be verified by microscopy (optical or electron, although optical detection allows *in situ* measurements). In addition, late-stage contamination by bacteria is noted by turbid medium (often described as "cloudy" or "milky" in appearance) and a sharp shift to lower pH, which registers as a color change in medium with a pH indicator such as phenol red. Mycoplasma do not scatter light as effectively as bacteria and are therefore difficult to detect by microscopy or turbidity measurements, and can remain undetected in the culture for long periods of time. Fungal contamination in the earliest stages can be characterized by presence of fungal bodies, as determined by microscopy. Growth of fungal bodies that are visible to the naked eye indicates late-stage contamination.

A contaminated culture, caught in the later stages, results in wasted reagents, cells, and time. If the sample is difficult to obtain (i.e., a primary-cell sample), then the cost of such a contamination is significant. Early-stage contamination, if undetected, may result in false results in an experiment, which may be difficult to trace if the sample is destroyed in the process.

Tip: It is advisable to always inspect cell samples under the microscope (at high magnification) to check for visible signs of bacterial or fungal contamination.

The relative cell concentration and health of the sample can be assessed using such a rapid and simple inspection prior to experimentation. When one considers the amount of time and cost associated with many cell analyses, there is little reason not to inspect each cell sample prior to analysis.

Contamination can occur from many places, including internal sources (reagents, other cell lines) and the external environment. If the cell sample is from a primary source (blood, tissue, etc.) then the contaminants from the donor source may carry over into the cultured sample. Also, it is almost impossible to remove tissue from a donor in a sterile environment, so the extraction/dissection process may also introduce opportunistic pathogens. Contamination of primary isolated cells, if used immediately after isolation, is not a significant problem. The exception, of course, is if the contamination was already affecting cell function *in vivo*. If the primary cells are to be cultured for any amount of time longer than a few hours, then antibiotics may be needed to maintain sterility. Loss of sterility may also result from the introduction of an unsterile reagent into the culture. Many dyes, antibodies, and labels are not screened for sterility, and introduction of these reagents negates any guarantee of sterility. Reagents that are ostensibly sterile, such as medium and sterile buffers, may also be contaminated unintentionally. A common cause of accidental contamination is when a reagent marked "sterile" is opened in the lab outside of a laminar biosafety hood, and then placed in the refrigerator or incubator without marking the bottle as "unsterile."

Tip: It is advisable, therefore, to segregate sterile and unsterile reagents and maintain strict labeling of each.

A good rule of thumb when dealing with cell cultures is "when in doubt, throw it away." The cost of a bottle of medium, while not necessarily trivial in all cases, is much smaller than the loss of cells and time from contamination. Considering the echoing effect of contaminated medium (i.e., several cultures possibly contaminated within minutes or hours of each other), the precaution of removing a suspect reagent from the lab is a sensible one. Serum, which is added to most cell cultures (Chapter 3), is a large source of contamination [1]. Bovine serum from a variety of vendors is guaranteed to be free from bacteria, fungi, and mycoplasma, but serum can (and in some cases should) be tested prior to use with particularly expensive or rare cultures.

Contamination can also occur from so-called external sources, such as cracked or compromised culture ware, or contamination from direct exposure to laboratory air or surfaces (i.e., not in a biosafety hood). Culture flasks should be inspected for cracks, particularly in the neck and cap. Petri dishes, which are usually stored in sterile packaging, must remain sealed prior to use to prevent accidental exposure to contaminants. While there are many points of entry for contaminants, performing cell handling in a sterile hood prevents many of these potential problems.

When dealing with a contamination, particularly a chronic one, careful isolation of each variable is needed. For example, a chronic contamination in one particular lab resulted in the loss of several cultures, six weeks of research time, and approximately $2000 (USD) in reagents and cell lines. After a lengthy isolation process, the source of contamination was determined to be an autopipettor (see Section 2.6 for discussion of these devices). After a thorough cleaning and sterilization of the autopipettor, all contaminations ceased. When a single Petri dish or culture flask is contaminated, it is most likely due to accidental exposure during the culturing operations (subculture, etc.). Petri dishes lack a definitive seal like culture flasks, and are therefore prone to contamination. When a second (or more) culture is contaminated within a close timeframe to the first contamination, then the reagents should be scrutinized for contamination. The medium, serum, buffers and reagents (if used), cells, culture ware, and liquid handling (pipettors, etc.) should all be tested methodically. Suspect medium should be placed in one or two sterile culture flasks, free of cells, and observed for several days in an incubator. If no bacteria or fungi are present, then the medium (and serum, if added) are not the source of contamination. If medium with serum is tested and found to be the source, the serum-free medium and the serum should be tested separately in a similar manner. Serum can be added to tryptic soy broth and incubated for 24 hours to test for contamination. In order to test buffers and reagents, they should be added to medium that has been determined to be sterile, and incubated. If none of the liquid reagents are determined to be the problem, then the pipettors should be cleaned and the process repeated with sterile medium.

Mycoplasma testing is more difficult and requires more advanced analyses than bacteria containing a cell wall. It is possible to test for mycoplasma using fluorescence microscopy and a DNA stain. Mycoplasma contamination appears as a fluorescent haze using optical microscopy of moderate (20–40×) magnification. However, this test should serve as a preliminary – rather than definitive – test, as background fluorescence from buffers and medium may also cause the same effect. At high optical magnification (100×) using immersion objectives (Chapter 4), mycoplasma may be observed as small (200–800 nm) fluorescent particles. Flow cytometry would also be possible under these conditions, although the scattering signals may be difficult to gate in a sample with a large concentration of debris. In fluorescence measurements, a Hoechst or other cell-permeant DNA dye is incubated in the cell sample, and the sample is inspected for sub-cellular particles (mycoplasma) [2]. Mycoplasma can also be detected by traditional microbial culture or immunological methods, as well as by polymerase chain reaction (PCR) assay. PCR and electron microscopy (Chapter 4) are definitive methods of detection, but may not always be accessible to all investigators. For those lacking the instrumentation or experience to conduct mycoplasma testing, many outside sources (e.g., American Type Culture Collection (ATCC) and Clongen Laboratories) offer mycoplasma testing and eradication services.

In addition to contamination by prokaryotic cells, a mammalian cell line may be contaminated by another mammalian culture. The sources of this type of contamination depend in part on the cell type. For primary cells, it is possible that cells from the same tissue, although of a different type, are present in the final sample. This is due largely to insufficient purification steps (Chapters 1 and 5). In some cases, it is nearly impossible to remove a background cell from a primary target cell culture. If the experiment can be conducted in the presence of the background cells, then this type of contamination is not a significant problem. If, however, the background cells affect experimental outcome, additional separation steps will be required. It is also possible for a cell sample, primary or immortalized, to become infected with another cell line. An example of this is the ATCC cell line CCL-13. This cell line was originally derived from liver tissue. Subsequent genetic analysis, however, indicated that over time this cell line was contaminated by HeLa cells (Chapter 1). Other cell lines have also been identified as HeLa contaminations, based on marker chromosomes from the HeLa parent line. While HeLa cell contamination represents the best examples of culture contamination by another cell line, this type of contamination is always possible.

Tip: As a precaution, only one cell sample should be manipulated in a biosafety hood at a time, so that reagents, flasks, and so on, are not accidentally mixed.

Since contamination of a culture by other cell types is difficult to detect, experiments can be affected for some time before a contamination is identified. There are commercial testing services available based on genetic identification of cells in a sample. It is also possible to test for cell contamination by flow cytometry, if one or more markers are identified as unique to the target or suspected contaminating cells.

2.3   SETTING UP A CELL CULTURE LABORATORY

Setting up a lab for cell culture is not an expensive task in the greater scheme of experimental research. There are some key pieces of equipment and supplies that are necessary for sterile culture work, regardless of the type of experimentation and analysis to be conducted afterward. Major cell-culture equipment can be categorized as either sterile-culture based or analysis/preparatory based. Culture-based equipment includes laminar biosafety hoods, incubators, and sterilizers. Other equipment necessary for cell handling and sample processing include centrifuges, culture ware, and personal protective equipment. Table 2.2 lists common items found in cell-culture laboratories and their uses.

Before any cells can be brought into a lab for continued use, both incubator(s) and laminar flow hood(s) must be present. A laminar flow hood, also called a biosafety cabinet (Figure 2.1a), ensures protection of both individual and sample or culture. Biosafety cabinets are rated for a particular biosafety level (BSL). BSL II hoods ensure that both sample and lab personnel are protected from each other, and are usually the best choice for most cell work. Biosafety cabinets use a sheet of laminar flow air to prevent airborne, aerosolized bacteria and fungi from entering a sterile cabinet. At the same time, aerosols generated inside the cabinet are not released into the laboratory. Figure 2.1b shows the airflow pattern in typical BSL I and BSL II biosafety cabinets. In the BSL I cabinet, laboratory air is free to enter the workspace, but contaminated air from aerosolized pathogens is contained in the cabinet. While the individual is isolated from the sample, it is possible for the sample itself to become contaminated. Therefore, BSL I cabinets are not recommended for sterile cell-culture work, but may be used in sample preparation (i.e., where the sample will

Table 2.2   Common equipment/supplies for sterile cell culture.[a]

	Use	Notes
Laminar Biosafety Cabinet/Hood	Culture/Handling	Serves as barrier between individual and cell cultures/ samples
Incubators	Culture	Sterile, homeostatic cell growth and maintenance
Liquid Nitrogen Dewar	Cell Storage	Large, multi-rack system for storage of cryopreserved cell lines and samples
Centrifuge (Medium size)	Sample Preparation	Sedimentation/pelleting of cells, separation of cells/ components, cell washing
Autopipettors (Serological)	Culture/Handling	Sterile fluid manipulation
Autoclave or H_2O_2 system	Sterilization	Preparation of culture materials
Culture ware	Culture/Handling	Petri dishes, culture flasks, and so on (sterile)
Microscope	Sample Preparation	Hemacytometer counting, cell inspection. Simple light microscopy adequate for these tasks
Water Bath	Culture/Sample Preparation	Warming of reagents to desired temperature. Thawing of cryopreserved cells
Refrigerator (4 °C)/Freezer (−20 °C and −80 °C)	Sample and Reagent Storage	Storage of medium, reagents, cryopreservation of cells (−80 °C)

[a] Excluding animal care.

eventually be introduced into an unsterile instrument). In BSL II cabinets, laminar air behind the front glass excludes lab air, while inside the workspace, a split airstream sweeps aerosols into the system intake. HEPA filters remove pathogen aerosols before the air is recirculated into the laboratory. BSL II type A cabinets are not designed for chemical hazards and should not be used with volatile or toxic compounds (unless in minute amounts for short durations). In addition, BSL II type A cabinets are generally not plumbed to outside ductwork for ventilation beyond the laboratory. BSL II type B cabinets are vented into the building hood ventilation system for chemical and biological protection. Class III cabinets are for BSL 4 organisms, which are not discussed in this book.

Biosafety cabinets, like any item exposed to the laboratory environment, are only as sterile as the individual makes them. There are general procedures for using a biosafety cabinet, although individual systems may have variations on the common theme. A general operating procedure for

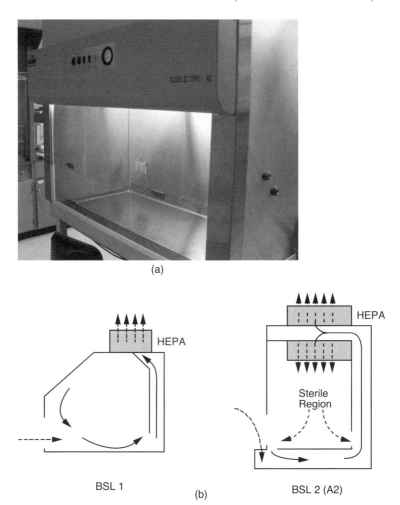

(a)

(b)

BSL 1

BSL 2 (A2)

Figure 2.1   (a) Photograph of a typical BSL II A2 biosafety cabinet. After sterilization, all work conducted behind the front sash is sterile; both the operator and the samples are protected. (b) Airflow diagrams of BSL I and BSL II A2 cabinets. Solid arrows indicate contaminated (sample) air; large-dashed lines indicated laboratory (unsterile) air; small-dashed lines indicate filtered air. BSL I cabinets protect the operator, but not the samples. BSL II cabinets protect both the operator and the sample, and are recommended for culture work or any time when sterility of the sample is important

using a BSL II type A cabinet includes system sterilization, reagent and sample delivery, and cleanup. Many biosafety cabinets require that the front sash be closed entirely before the UV light is turned on to sterilize the cabinet surfaces. If pipettes or other culture ware are to be used, it is possible to sterilize the surface of these items and their packaging by placing them inside the biosafety cabinet before the UV light is turned on. It is advisable to anticipate the number of items needed so that the biosafety cabinet is not crowded and airflow disrupted. A 5–10 minute UV exposure will sterilize the cabinet and items placed inside of it. Medium, cell samples, and any reagents should not be placed in the biosafety cabinet during UV sterilization – this step is only for empty cultureware items and the cabinet surfaces. The sash is then once again placed at the proper height while the blower is operated. In some cabinets the blower motor is tied to the sash-height sensors so that the blower shuts off when the sash is down. If this is not the case, then the blower must be manually turned on and off as the sash is raised or lowered to avoid motor damage. Before a hood is used, the flow rates should be checked (this can be performed as periodic maintenance) to assure proper laminar flow. The front glass sash should be kept at the appropriate height as determined during the periodic cabinet inspection. At this point any cells, medium, and reagents can be introduced into the cabinet for use.

Some protocols specify that bottle tops or other items should be flamed using an open-air Bunsen burner (or similar) to sterilize them before opening in the biosafety cabinet. However, many biosafety cabinets explicitly prohibit the use of flames inside the cabinet; indeed, the associated safety issues make other sterilization methods more attractive. Spraying down the bottle necks with 70% ethanol is effective for removing any outside contaminants before fluid handling occurs. Specific operator protocols and advice for handling cultures and reagents in the hood are discussed in Chapter 3. It is best to work in batches, handling only one cell type at a time, and avoiding building up reagents, bottles, and other items that obstruct airflow in the cabinet. Once all work is completed, the cells are returned to the incubator or processed in the lab if sterility is no longer required. All reagents and culture ware should then be removed from the hood and the work surface cleaned with 70% ethanol and wiped down with a paper towel or lab wipe. The paper towel or wipe is then disposed of in biohazard waste. At this point, the hood can be left operating, can be shut down (and the sash lowered), or sterilized again for a different cell type.

2.3.1   Tissue Culture Incubators

The primary function of tissue culture incubators is to maintain homeostasis – primarily temperature – of cultures and samples. The incubator itself is inherently an unsterile environment; the culture ware is the barrier to bacteria and fungi. Culture incubators vary by design, but all provide a level of temperature and atmospheric control. For mammalian cell culture, water-jacketed incubators provide excellent temperature stability due to the large thermal mass of the surrounding water layer. Many incubators also feature a two-door system, with an internal glass or polycarbonate door to view cultures without disrupting the atmospheric equilibrium. The outer door ensures temperature stability. Each time the inner door is opened, the incubator air is mixed with laboratory air of lower temperature, humidity, and CO_2 content. Incubators are operated at $37\,^{\circ}C$, 90% relative humidity, and 5% CO_2 for most mammalian cell cultures. Humidity control in higher-end systems is achieved using a steam generator. In most incubators, high humidity is established using a reservoir of water at the bottom of the incubator that must be refilled periodically. While convenient, this water reservoir is a prime source of many of the bacteria and fungi in the incubator itself, and should be cleaned often. The incubator walls can be wiped down with dilute bleach or a cleaning agent recommended by the manufacturer.

An atomosphere of 5% CO_2 is required for medium containing carbonate buffer, to maintain pH. CO_2 is supplied by an external tank and the gas supply line is usually fitted with a $0.2\,\mu m$ filter to reduce contamination and particulates from the CO_2 tank. Only gaseous compressed CO_2 cylinders should be used. If the tank is overfilled and liquid CO_2 exits the cylinder, then the gas regulator, gas lines and filter, and the incubator itself can be damaged. It is recommended that each CO_2 cylinder is tested outside (secured to a tank cart or solid support) to ensure that no liquid CO_2 escapes.

When using an incubator for multiple cell lines, which is often the case, they should each reside in a designated section of the incubator. Doing so will reduce the likelihood of cross contamination of cell lines. Petri dishes and nonsealed culture chambers should be segregated from culture flasks and their storage area cleaned more frequently, as they are prone to spilling. If bacteria are also to be cultured, it is advisable to have two culture incubators, one for prokaryotes and one for eukaryotes.

2.4   CELL LINE STORAGE

Immortalized and primary cells can both be cryopreseved and stored for future use. In the case of immortalized cells, it is advisable to split a culture into several cryovials so that backup cultures can be made in case of contamination, incubator power failure, or other accidents that result in loss of a culture. In some cases, cell lines obtained from outside sources cannot be stored in this manner due to liscensing reasons; the agreement for each cell line will state if cryogenic storage is prohibited. For primary cells, storage of multiple cell samples from the same animal or tissue is recommended to ensure reproducible measurements. However, the cryo-preserved samples will only be useable if the storage and thawing process does not alter cell function.

Chapter 3 discusses the protocols and considerations of cryopreserva-tion. However, for final storage, a liquid nitrogen Dewar or $-80\,^{\circ}$C freezer is needed. Freezer storage poses a risk if power is lost for prolonged periods. Using a liquid nitrogen Dewar requires periodic filling of liquid nitrogen, but is otherwise unpowered and safe from loss of cell samples. Liquid nitrogen (LN_2) Dewars for cell and sample storage feature double-walled vacuum insulation and racks for storing cells (Figure 2.2a). Unlike smaller LN_2 Dewars designed for transport of liquid nitrogen, these dewars have a constant amount of LN_2 at the bottom to maintain temperature. Typically, only a few inches of LN_2 are kept at the bottom so that the LN_2 vapor phase is responsible for temperature

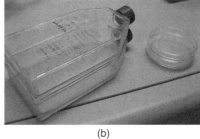

(a)   (b)

Figure 2.2   (a) Photograph of a medium-sized liquid nitrogen (LN_2) Dewar for cell and sample storage. Each rack (shown being removed from the Dewar) contains several sample boxes, which each in turn house several sample cryovials. (b) Examples of culture ware, such as culture flasks (also called T flasks) and culture dishes)

regulation. There are several reasons not to store cultures directly in the liquid phase. First, it is possible for cross-contamination to occur between vials stored directly in liquid nitrogen. Second, the liquid can directly enter the cryovial, creating an explosion hazard when the vial is removed and the liquid nitrogen expands. For this reason, the vial can be briefly "vented" in the biosafety cabinet after removal. The temperature in the vapor phase is not as stable as direct liquid immersion, but even at the relatively higher temperatures cell lines remain frozen. LN_2 vapor-phase storage requires less LN_2 than direct immersion in the liquid and is potentially safer as there is less LN_2 to spill.

The LN_2 Dewar will continue to properly store cells as long as the liquid nitrogen level is maintained. A weekly check using a wooden meter stick will ascertain the level (a hollow measuring device should never be used, as LN_2 can be aspirated up through the tube). The main safety issues with LN_2 Dewar storage are cryodamage to exposed skin and the danger of displacement of lab air with gaseous nitrogen. The former concern can be minimized if those handling cultures wear the correct personal protective equipment. Liquid nitrogen Dewars used in small laboratories, storage spaces, or other enclosed areas should be fitted with oxygen sensors with alarm capabilities. The high expansion rate of liquid nitrogen can create a displacement of lab oxygen, resulting in asphyxiation. If a lab is discovered with an alarming oxygen sensor, proper breathing apparatus must be worn before anyone enters the room. This precaution must be followed even if an individual is found incapacitated in the room. With smaller amounts of LN_2 in larger laboratories, an oxygen sensor may not be needed; it is best to consult with the institutional safety personnel to determine if a sensor is required. However, given the relatively low cost of oxygen sensors ($300–$900 USD), outfitting a laboratory with one is an inexpensive precaution.

Liquid nitrogen Dewars maintain their temperature as long as the system is opened minimally when retrieving cryovials. Therefore it is important that the exact location of the vial in question is known. A log book containing each cell sample, the individual who cryopreserved it, and its location in the Dewar is necessary and often required for safety regulations. A typical LN_2 Dewar will have several racks that each contain several cryovial boxes, which each box containing at least 25 vials. A log book should therefore identify the Dewar (if more than one is used), the rack, box, and position of each cell line. Another issue with storage and retrieval of cryopreserved samples is that all labeling must be LN_2 compatible. A cryovial that has lost its label must either undergo expensive genetic testing for identification, or be discarded.

2.5   PERSONAL PROTECTIVE EQUIPMENT

Along with the biosafety cabinet, personnel protective equipment (PPE) items are the best protection for individuals working in a cell laboratory. A distinction needs to be made between chemical and biological safety, although the two are usually closely linked in cell analysis. Both chemical and biological PPE require proper footwear and clothing to risk exposure to skin and damage from sharp objects that may be dropped or fall off of a lab bench. Most chemical laboratories require lab eyewear, and most cell labs should require the same. Biological safety PPE extends to additional coverings and, most importantly, strict use of gloves to protect the hands. There are two main types of lab coats found in cell-culture labs, both intended to protect individuals and their clothing from spills, and so on. Open-cuffed lab coats are appropriate for general lab use outside of the biosafety cabinet. For work in a biosafety cabinet, elastic-cuffed lab coats add additional coverage and protection for sterile work.

Gloves should be worn when handling any cells, samples, or reagents. It is also important to *not* wear gloves when using telephones, keyboards, and so on in the lab. These items should be universally regarded as "cell- and reagent-free" areas. Glove materials vary, but most for cell work are either latex or nitrile. Latex gloves are used for most clinical work, while nitrile gloves have superior chemical resistance. Both glove types can be used for cell culture. Gloves can be sterilized in an autoclave or hydrogen peroxide (H_2O_2) system, but for most culture work, gloves can be sterilized in the hood using 70% ethanol. Gloves are always single-use items and should be discarded in biohazardous waste when finished. In addition to the standard lab gloves for handling cells and reagents, specialty gloves may also be needed. For example, anti-needle-stick gloves can be used to reduce – but not eliminate – the possibility of needle sticks. There are several manufactures of these types of anti-syringe gloves, such as the Stichstop from KCL or the series of gloves from HexArmor. These gloves can be worn over nitrile or latex gloves and are reusable in some cases. They often feature a mesh design to reduce the risk of puncture, and can also be used to avoid cuts from other sharp objects. Anti-syringe gloves do reduce the dexterity of the user, and are not always suitable for the most delicate of tasks. If needles are handled routinely, such as in disposal of biohazardous waste, then these gloves are a useful investment. There are also a range of other technologies that have been introduced to eliminate needle sticks. Self-sheathing needles, which retract upon insertion and are covered by a shield upon withdrawal, reduce the risk of accidental punctures. Blunt canulas are also an alternative to needles, although they cannot be used for direct subcutaneous or intravenous injection. Canulas can be used with injection ports, which can be fitted to a variety of tubing connections and devices.

In the culture lab, face shields are also required when there is a risk of a splash hazard. For example, if a cryovial is to be removed from liquid nitrogen vapor-phase storage, there is a risk that trapped LN_2 can rupture the vial and disperse its contents. A face shield should always be worn during sample storage and retrieval from LN_2 Dewars. Face shields should also be worn when refilling LN_2 Dewars. Other times when face shields should be worn include centrifuge servicing, and cleaning of accidental sample spills (to avoid any splashing from cleaning, even if bleach or ethanol is used).

2.6   CELL AND SAMPLE HANDLING

During routine maintenance of cell culture or preparing cells for analysis, the samples must be handled in a sterile manner that also preserves viability and cell function. Therefore, most cell handling, if not all, must be conducted in a biosafety cabinet. The main mechanism for handling cells in suspension is pipetting. Other methods, such as pouring, are not recommended, as they are both imprecise and prone to introducing contaminating organisms. While suspended cells can be handled without additional reagents, adherent cells must be treated to be released into suspension (Chapter 3).

Even if the cells are not moved before analysis (i.e., if they are grown directly onto a coverslip or microscope-compatible dish, Chapter 4), some form of liquid transfer or introduction of reagents is required. For small volumes, pipettors with disposable tips should be used. Sterilized pipette tips are needed if the reagents and cells are to remain sterile. For volumes larger than one milliliter (ml), serological pipettes are recommended. The sterile, individually packaged serological pipettes are available in a variety of volumes and are typically graduated. Bulk packaged serological pipettes are prone to loss of sterility. Serological pipettes should not be operated using a suction bulb; instead, an automatic pipettor should be used (Figure 2.3). Most automatic pipettors use 0.22 μm barrier filters to prevent accidental aspiration of culture into the pipettor, reducing the risk of cross-contamination.

Items that are needed for sterile culture, but are not sterile (or need to be resterilized), must be subjected to some form of sterilization. Microfluidic devices require some form of direct sterilization using a sterilizing solution for the internal fluid circuits. Larger items can be placed in an autoclave or

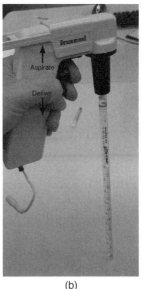

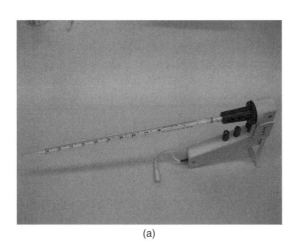

(a)   (b)

Figure 2.3   (a) An autopipettor capable of aspirating and dispensing fluid (shown with a single use serological pipette inserted). (b) Aspirate and dispense (deliver) controls of the autopipettor

H_2O_2 sterilizer. Autoclaves are adequate for most culture work to sterilize gloves, glassware, pipettes, and other items. Any item placed in the autoclave must be resistant to the pressures and temperatures used in an autoclave. Items that will melt, compress, or deform in the autoclave, or items that are water sensitive, can be sterilized in a H_2O_2 sterilizer. H_2O_2 sterilizers are less common than autoclaves, but can typically be found at most hospitals. H_2O_2 sterilizers can only be used to sterilize dry materials; items containing water or water vapor must be sterilized by autoclave. To ensure sterility after processing, items should be carefully wrapped in autoclave sheets or – preferably – sealed in sterilization bags. Adhesive labels indicating proper sterilization can be added as a safety measure to ensure sterile products. Items sterilized in autoclave bags can be stored outside of the culture hood until needed.

2.7   COMMON ANALYTICAL INSTRUMENTATION FOR CELL CULTURE

While the scope of this book is to provide practical advice on a variety of cell analyses, there is a sub-set of instrumentation that is essential for cell culture itself. The most important perhaps is a light microscope. Light (transmission) microscopes and other microscope types are discussed in Chapter 4 in detail. A light microscope capable of suitable magnification ($40\times-400\times$ total magnification) to observe cell cultures is critical in many aspects of cell maintenance. Therefore, a typical microscope with $10\times$ eyepieces and $4\times-40\times$ objectives will suffice. A brief observation of a culture flask can indicate approximate cell density and overall cell health. For sub-culturing adherent cells (Chapter 3), the same microscope will indicate when cells have been released from the surface. If the microscope has fluorescence capabilities, then a short visual observation will confirm staining of many dyes (although typically not dim cells that are stained by a small number of dye molecules). From a preparatory standpoint, one of the most compelling reasons to keep a light microscope for cell culture is the measurement of cell concentration using a hemacytometer. A hemacytometer (Figure 2.4), is a ruled glass device designed with a finite volume for counting cell types. It is used in conjunction with a microscope to determine cell concentration. Since many staining protocols require a certain concentration range, a hemacytometer counting step is often necessary. In addition, methods that only offer relative cell numbers, such as most flow cytometers, can be corrected and expressed in units of true concentration (cells ml^{-1} etc.). While counting beads can be used to calibrate relative cell-count analyses, a hemacytometer can achieve the same result, although it becomes tedious for many cell samples.

The hemacytometer image shown in Figure 2.4 shows both the device and a typical field of view for white-blood-cell counting. The hemacytometer in Figure 2.4 uses Neubauer ruling to divide the $9\,mm^2$ area. Figure 2.4b shows a conceptual diagram of the hemacytometer grid. The outer four squares ($1\,mm^2$ each) that occupy the grid corners are divided into $0.25\,mm \times 0.25\,mm$ sections. The depth of each $0.25 \times 0.25\,mm$ section is $0.1\,mm$ below the cover glass, yielding a volume of $6.25 \times 10^{-3}\,mm^3$ ($1\,\mu l = 1\,mm^3$). This region is suitable for white-blood-cell counts or other large cells such as cancer cultured cell lines. The central square is divided into 25 smaller sections, each containing 16 squares. These 16 squares have vertical and horizontal dimensions of $0.05\,mm$ each, resulting in a volume of $2.5 \times 10^{-4}\,mm^3$. Cells are counted in each grid (either the central grids or the four corner grids) and the final concentration is calculated.

The center square millimeter is used to count erythrocytes and bacteria, as well as small microparticles. While magnification can vary, count cells using a $20-40\times$ objective. It is best to count at least four squares to obtain a suitable average. The cell sample may have to be diluted if the field of

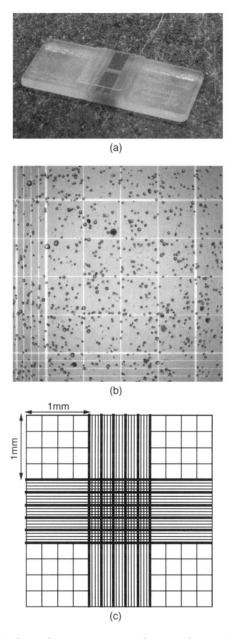

Figure 2.4   (a) A Neubauer hemacytometer with cover glass, suitable for white-light transmission measurements. (b) Image of acute leukemia lymphocytes on a 1 mm² section of the hemacytometer. Each square inside the triple-lined perimeter is 0.25 mm 0.25 mm × 0.1 mm (l × w × h). (c) Diagram of Neubauer ruling, showing the four corner squares used for leukocytes. The center square ruling is used for erythrocytes, bacteria, and other cells of the same size scale

view is too crowded. If the cell sample is too sparse, then more sample squares should be counted (both sides of the hemacytometer can be used with the same sample if more squares are needed). For this example, assume 11 red blood cells were counted in ten $0.05 \times 0.05 \, \text{mm}$ squares. The total volume of these ten squares is 10 $(2.5 \times 10^{-4} \, \text{mm}^3)$ $2.5 \times 10^{-3} \, \text{mm}^3$ or μl. If 11 cells were counted in total over those 10 squares, the cell concentration is 11 cells per $2.5 \times 10^{-3} \, \text{μl} = 4400 \, \text{cells} \, \text{μl}^{-1}$ or $4.4 \times 10^6 \, \text{cells} \, \text{ml}^{-1}$. If any dilution factor was used, such as a twofold dilution, then that factor must be used to correct for the original concentration. For example, if the sample was diluted twofold before hemacytometer counting, then the original concentration was $2 \times (4.4 \times 10^6 \, \text{cells} \, \text{ml}^{-1}) = 8.8 \times 10^6 \, \text{cells} \, \text{ml}^{-1}$.

For white blood cells and other, larger cell types, the corner grids are used. Each corner square millimeter has 16 sub-divisions with a volume of $6.25 \times 10^{-3} \, \text{mm}^3$ each. The sum of the number of cells is calculated as the previous example. If 47 cells are counted over four of the $0.25 \times 0.25 \, \text{mm}$ squares (total volume of $0.025 \, \text{mm}^3$), the cell concentration is 47/ $(0.025 \, \text{mm}^3) = 1880 \, \text{cells} \, \text{μl}^{-1}$. Again, any dilution factor must be added to adjust the value to the original concentration.

There are variations of this most common form of hemacytometer, including those designed for fluorescence. Fluorescence or standard, white-light hemacytometers can be used to perform simultaneous counting and viability measurements. A membrane-impermeant dye suitable for fluorescence or white-light imaging can be used to identify dead and live cells during hemacytometer counting (a two-channel tally counter is useful for this task).

If specimens are dissected, then a stereomicroscope is also a recommended item in a culture lab. Typically less expensive than transmission and fluorescence microscopes, stereomicroscopes are well suited to guiding the manipulation of small objects, such as connections of microfluidic interconnects or handling smaller animals or organs. Stereomicroscopes are either two compound microscopes that share a common focal plane, or a common focal optics set that is split into two ocular paths. In the case of the former, the two compound microscope systems are at different angles to form the stereo image. Stereomicroscopes are almost always upright systems with moderate magnification and a larger working distance suitable for manual sample manipulation. Lower-end stereomicroscopes often suffer from distortion from the diverging optical paths. More expensive systems correct for these image distortions, and should be used when high spatial resolution is needed. For most cell, sample, or microfluidic device work, a mid-range stereomicroscope is sufficient.

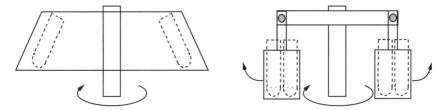

Figure 2.5 Cross-sections of a fixed rotor and a swing-out rotor for centrifuges. Dotted lines represent the placement of sample tubes. Fixed rotors can achieve high speeds, but are limited in sample tube size. Swing-out rotors can accommodate many different cell types, including multi-well plates

Centrifuges are another essential piece of equipment, regardless of the type of cell analysis being conducted in a laboratory. For most cell preparation, particularly for suspended cells, a centrifuge capable of handling 0.5–1.5 ml centrifuge tubes is needed. Centrifuge tubes for blood samples (12×75 mm) are also needed for many flow cytometry preparations. For larger cell volumes, 50 ml tubes are needed, particularly for freezing entire cell culture flasks (Chapter 3) or concentrating a dilute cell sample for analysis. For most scientists, it is advantageous to invest in a centrifuge system that will accommodate different size rotors so that different tubes and sample types can be used with one piece of equipment. Rotors for centrifuges are configured as fixed angle or swing-out buckets (Figure 2.5). The former achieve higher speeds and can accommodate many sample tubes, but the swing-out system allows for nontraditional sample containers to be centrifuged (e.g., 96-well plates or similar). Refrigerated centrifuges are recommended for samples that require thermostatting or for long-duration centrifugation steps.

The speed (and centrifugal force) should be estimated for the desired separation. Chapter 9 describes the conversion of the gravitational force and the speed (revolutions per minute) of the centrifuge rotor. As many protocols list the relative centrifugal force, and many centrifuges read out only the rotor speed, it is important to perform this calculation in many cases.

2.8 CONSIDERATIONS WHEN SETTING UP A CELL-CULTURE LABORATORY

When dedicating space in an existing lab, or setting up a new research effort, the location and layout of a culture lab is critical for safety, sterility, and efficiency. In general, it is better to have a culture lab that is isolated from other laboratories. Doing so will prevent contamination of other equipment and to keep the culture area as aseptic as possible. A culture laboratory must house all of the necessary equipment and also allow for a safe and productive workflow as individuals remove cultures from incubators, manipulate them in biosafety cabinets, and prepare them for analysis. It is not necessary to house instruments that do not preserve sample sterility in the same facility, although a light microscope should be located as close as possible to the culture room. Figure 2.6 depicts two floor diagrams of successful culture laboratories each approximately 400–500 ft^2 (37–46 m^2); they are by no means the only permutation or even a recommendation for setting up a culture room. However, they do serve as an example to demonstrate the ability to place the necessary components in a relatively small area.

In Figure 2.6a, the laboratory in question is a space dedicated to cell culture, cell preparation, and some minimal microfluidic-device preparation. Not shown in the figure are chemical hoods that are not used for culture purposes. The culture area, consisting of the laminar flow hood, the incubator, and refrigerator, are all located within reach of the individual using the biosafety cabinet. This arrangement allows reagents or culture flasks to be retrieved easily while the individual remains seated. The lab chair for cell culture can be made of plastic of leather, or covered if it is upholstered with porous materials for easier cleanup of spills. The unsterile preparation areas are located on benches along the walls, well away from the cell-culture area. The clustering of sterile and unsterile work space minimizes contamination and also allows for an ergonomic workflow. Figure 2.6b shows a slightly larger lab that was used for cell culture as well as cell analysis. In this case, the geometric limitations of the facility did not permit as efficient a workspace. However, the culture region and incubators are located in close proximity. Some equipment, such as water baths and LN$_2$ Dewars, were located in a different room. These two example laboratories differ in layout, but feature common items and designs to maximize the utilization of space and minimize contamination risk.

Setting up a laboratory initially requires some thought about workflow, the type of work that will be conducted, and the limitations of the investigator's physical space. Maintaining that laboratory, however, requires adequate funding to keep a continuous stock of reagents, culture ware, and other consumables required for culture and analysis. Most items that will contact a cell or reagent in some way are single use, disposable items. The use of single-use items reduces contamination and cross-contamination risk. The added benefit for vendors is the guaranteed

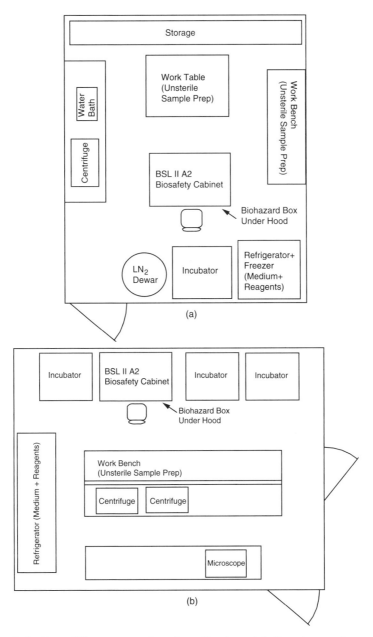

Figure 2.6 Two different examples of a cell-culture lab floor layout. In general, biosafety cabinets and incubators are kept close together for ease of workflow. Other equipment, such as microscopes and centrifuges, can be kept away from the sterile work area

Table 2.3   Common expendable items for cell culture and analysis

Item	Cost	Application	Use
Culture Medium	Moderate	Culture	Multiple
Serum	High	Culture	Multiple
Serological Pipette	Low	Sterile Handling	Single
Micropipette Tips	Low	Liquid Handling	Single
Microscope Slides	Low	Analysis	Single
Microfluidic Devices	Moderate-High[a]	Analysis	Single[a]
Syringes (Sterile)	Low	Analysis	Single
Gloves	Low	Multiple	Single
Cytometer Tubes	Low	Flow Cytometry	Single
Antibodies	High	Analysis	Multiple
Dyes/Probes	Moderate-High	Analysis	Multiple
Culture Reagents (proteases, etc.)	Moderate	Culture	Multiple
Conjugation Reagents	Moderate	Analysis	Multiple
Buffers	Low-Moderate	Culture Analysis	Multiple
Microscope Lamps	Moderate	Analysis	Multiple

[a] The assumption is that most microfluidic devices are still made in the lab, rather than purchased, so the cost and number of uses is variable.

repeat business by cell analysis researchers needing to restock their labs. Table 2.3 lists some of the consumable supplies that are used routinely in cell culture and preparation for many of the analyses discussed in this book. Table 2.3 does not consider animal-dissection supplies, which increase the cost. Rather than list costs of individual items, which are always in flux (and always on the rise), it would be more beneficial to recommend that each investigator track consumables expenditures for a year to estimate the monthly cost. In one particular laboratory, expenditures for consumables alone was $1000–2000 USD for a research group of 8–12 people. Keeping track of these types of fund requirements strengthens arguments for budgets in grant proposals, and also allows the investigator to more effectively manage funds.

The discussion of cell-culture laboratories and equipment so far has focused on the culture and maintenance of cell lines. If primary cells are to be isolated from donor animals, additional facilities for animal care are needed. Each institution will have established animal-care protocols and requirements, often with published costs associated with animal care. These guidelines and figures allow investigators to estimate costs of adding animal care to their research capabilities; these values are often listed per animal per day. If animal care or tissue harvesting occurs in another lab, then the setup of the cell lab is similar to one designated for secondary (immortalized) culture only.

2.9   ESTABLISHING AND REGULATING A CULTURE FACILITY

When setting up a new culture facility, it is advisable to work closely with institutional safety officials and CDC guidelines. Doing so from the onset will ensure proper lab operation, safety, and minimized risk of sample contamination. It is also critical to establish proper conduct and protocols before the investigator is cited for improper procedures. Safety inspections will include annual maintenance and measurement of biosafety cabinets for proper airflow. This yearly inspection ensures proper isolation of sample and individual, and is often required for a lab to pass inspection. The inspection requirements are usually scaled by what types of specimens are worked with in the lab (i.e., BSL 1 vs. BSL 2, etc.). Additional inspection points include sample storage (refrigerator, incubator, LN_2 Dewar), and general cleanliness of the workspace, data logs, and waste disposal (both chemical and biological).

A properly regulated cell-culture lab begins at the entryway to the room. Are proper signs in place? One particular institution required all cell types/ lines to be listed on the door, along with their BSL rating. Another institution only required the BSL rating for the entire facility. In both cases, proper biohazard signs (see Figure 2.7) were required to warn outside individuals of the risks of entering. The use of appropriate signage is especially important for maintenance staff, so that they remain protected while working in the room. If personal protective equipment is required, then any custodial or maintenance staff should be provided with them before they enter the room. If a special license is required – as is often the case for BSL 2 work – then a copy of that license and emergency contacts should also be placed on the door. Unless it is required by institution regulations, it is not advisable to label the outer lab doors with signage indicating animals are housed in the facility. While most animal research debate remains civil, there are instances in which laboratories have been broken into and vandalized. While a discussion of the ethics of animal research is beyond the scope of this book, it is recommended that if such signage is required, additional security measures should be implemented to reduce the risk of vandalism.

One of the most important aspects of regulating a cell-culture lab is the proper storage and disposal of biohazardous waste. While regulations vary by institution, generally anything that has come in contact with cell or tissue samples, medium, serum, or anything else that can harbor infectious agents should be disposed of in biohazardous waste. In the

Figure 2.7 Example of a biohazard door sign (for use in the United States). Additional signage, such as cells/tissues housed in the laboratory, may be required. Institutional safety committees will have specific requirements for door warnings and labels

United States, disposal of sharps (needles, etc.) is performed in specialized containers to reduce the risk of puncture during waste removal. Other waste, such as culture flasks, serological pipettes, gloves, and other items can be stored in a biohazard box or container lined with the appropriate biohazard waste bags. These bags minimize the risk of leaks and are autoclavable. In some institutions, cardboard biohazard boxes are used, which are subsequently incinerated. The bag can be removed (e.g., if a plastic receptacle is used) and in many cases incinerated as well. Alternatively, the biohazard bag can be removed, autoclaved, and then disposed of. If animal carcasses or tissue are involved, then a double-barrier system (an inner and outer bag) should be used to transport the material to the disposal/incineration site. The regulation of infectious waste should be anticipated, as a plan of disposal is often required for IACUC approval.

2.10   CONCLUSION

While cell and tissue research is varied in scope and approach, there are common themes that can be used to set up a functional cell-culture laboratory. Establishing a culture facility is not a task that requires an extraordinary budget or timeframe. The most expensive components (excluding analytical equipment such as microscopes) are biosafety cabinets, centrifuges, and incubators. As all three items are essential for any type of cell work, they are valid investments. Individuals in a cell-culture lab must be constantly vigilant to avoid contamination or safety risks. When accidents do happen, the incident must be reported and all institutional guidelines followed. In the case of contamination, careful isolation of all variables can determine the source of contamination and minimize future risk. Like any lab, a culture facility requires thorough supervision, regulated procedures, and a clean, ergonomic work area. Spending time and effort on the initial planning and implementation of a cell culture lab will provide a facility that will enable diverse and challenging research in the future.

REFERENCES

1. McGarrity, G.J., Gammon, L., and Coriell, L.L. (1980) Detection of *Mycoplasma hyorhinis* infection in cell repository cultures. *Cytogenic and Genome Research*, **27**, 194–196.
2. Battaglia, M., Pozzi, D., Grimaldi, S., and Parasassi, T. (1994) Hoechst-33258 staining for detecting mycoplasma contamination in cell-cultures – a method for reducing fluorescence photobleaching. *Biotechnic and Histochemistry*, **69**, 152–156.

3

Maintaining Cultures

3.1 INTRODUCTION

Chapters 1 and 2 dealt with the procurement of cells and the establishment of a cell-culture laboratory, respectively. With the starting materials and tools now in place, this chapter discusses the maintenance of cells for bioanalysis. The proper maintenance of cells includes homeostasis during culture, cell storage, and the proper preparation of cells for analysis. The latter case is most important, as often analysis and homeostasis are incongruent. Buffers must be changed, different media used, and the cells, at times, are exposed to drastically different conditions for analysis. In some cases, the change in conditions can affect the outcome of the experiment negatively. In other instances, the conditions proper for cell analysis are fatal to the cell (e.g., electron microscopy).

There are many texts available on the culture of almost every cell type imaginable. This chapter focuses on the maintenance of cells for analysis, and the proper analysis of cells. Various supplements, and their affect on cell culture and analysis, are discussed. The effects of sample handling, cryopreservation, and two- or three-dimensional culture are discussed from a bioanalytical standpoint. When culturing primary or immortal cells for analysis, sterility and cross-contamination must also be monitored at all times. A few bacteria in a sample can wreak havoc in a short time, rendering any analytical data useless. The cross-contamination of cultures (see Chapter 2) is at best a nightmare, as extensive genetic testing is required to purify cell populations and yield accurate data.

Practical Cell Analysis   Dimitri Pappas
© 2010 John Wiley & Sons, Ltd

Considering the cost of cells, reagents, instrumentation, and lab up-keep, at least as much thought should be placed on the maintenance of cell cultures for proper analysis. The type of environment the cell encounters can directly affect the outcome of an analytical experiment. Cell-growth conditions, analysis buffers, and reagents can affect cell phenotype, cell signaling, and a host of other parameters. By careful maintenance of primary and immortal cells, accurate and reproducible cell analyses can be conducted.

3.2   MEDIUM

More than any other reagent in a cell-analysis laboratory, a steady supply of culture medium – and the choice of correct medium type – is essential for cell analysis. There are, in general, two classes of medium one can consider for cell analysis. First, medium that is used to maintain a culture in between experiments, and second, medium used in the analysis itself. Often these two can be one in the same, although in some cases a modified medium or supplemented buffer is needed during the analysis or processing phase. There are many types of medium available, and the supplements that can be added to them expand the palette of options even further. Table 3.1 lists some medium types that are common to cellular analysis, by cell type. The table is not inclusive, but serves to highlight the differences in medium types, and that some medium formulations are applicable to many cell lines. In most cases, the medium in Table 3.1 is used during the culture (maintenance) phase, and a different buffer or medium may be used during the analysis itself.

Medium can be classified as basic or complete, depending on whether or not serum is included, respectively. Basic medium has many of the components required for cell metabolism. Basic media, such as DMEM and RPMI 1640 (see Table 3.1), contain salts (partly from buffer action), amino acids, vitamins (such as biotin, folic acid, B-12, etc.), and molecules involved in energy production (glucose, pyruvate). Basic medium also often contains other buffers (such as HEPES) and a colorimetric acid–base indicator, such as phenol red. The latter serves as a quick visual inspection of the "age" of the medium in culture. As cells consume nutrients and produce waste, the culture medium acidifies, resulting in a shift in color for the pH indicator. The formulations of most culture media are available and should be examined for potential interference in analysis. For example, staining using Annexin-V-based apoptosis probes requires relatively high Ca^{2+} concentrations. At the same time, the presence of

Table 3.1   Medium types common to cell analysis

Medium [ref]	Serum	Additives	Cell lines[a]
RPMI 1640 [1–5]	10% FBS	Antibacterial–Antifungal	Jurkat, HuT 78, RPMI 8226, CCRF-CEM, U937, HL-60
Dulbecco's modified Eagle Medium (DMEM, [6–8])	10% FBS	Antibacterial–Antifungal, L-Glutamine	NIH 3T3, RBL-1, HT-29, HeLa
Claycomb's Medium [9]	10% FBS (Special Lot)	Antibacterial–Antifungal, Norepinephrine, L-Glutamine	HL-1
Cell Mab [10]	0%–10% FBS	Varies	Designed for monoclonal antibody production
Leibovitz's L-15 [11]	Hemolymph		Bag neuronal cells
Eagle's Minimum Essential Medium	0–10% FBS	L-Glutamine	
F-12	0–10% FBS	L-Glutamine	Designed for primary cells
Iscove's Modified DMEM	0–10% FBS	L-Glutamine	HuT 78 T Cells

[a] See Chapter 1.
FBS = Fetal Bovine Serum.

phenol red in the medium will interfere with fluorescence measurements of fluorescein, green fluorescent protein (GFP), and other fluorophores with similar emission spectra. As discussed in Chapter 4, fluorescence from phenol red itself makes sensitive fluorescence measurements nearly impossible.

Medium is, in essence, a man-made attempt to mimic the life support found *in vivo*. It is, therefore, lacking in many essential compounds for cell growth. Many cell lines can function in basic medium without additional materials, but for most routine culture and analysis, serum must be added to form the complete medium. Serum is typically derived from animal sources, the most common being fetal bovine serum (FBS). FBS and other sera contain growth factors such as epidermal growth factor (EGF), some interleukins, and transferrin. Also present are adhesion-promoting proteins and peptides such as fibronectin and laminin. Other components include insulin and various minerals. FBS and other animal-based sera are by far the most common supplements used for culture maintenance. Being derived from animal sources, serum is inherently difficult to use from a quality-control perspective. Serum derived from different animal types can affect experiment outcome [12]. For example, the use of FBS instead of native rat serum was shown to affect the outcome of rat leukocyte immunological response. In addition to species variability, serum varies from lot to lot, as well as by country of origin. If cell products are to be analyzed over long time periods (months of experimentation) it is best to purchase a large quantity of serum from one particular lot. Given the high cost of medium ($200–$1000 USD per liter), this may not always be practical. Serum cost increases as the level of quality control improves. The more consistent and well characterized the medium, the higher the cost.

Another negative aspect of dealing with serum is that the serum, or animal of origin, is subject to contamination, just like any other primary-derived material. Certain viruses, bacteria, and mycoplasma have been shown to be transmitted via serum (Chapter 2). There are several replacement sera that can be substituted for FBS. For example, the FetalClone series and Bovine Growth Serum, both from HyClone, are non-fetal animal sera supplemented with various growth factors, minerals, and other compounds. Since they are not derived from fetal animals there is less variability between lots (especially for the added compounds). None of the alternative sera offer much relief as far as cost is concerned, but the increase in quality control is a major improvement.

Some cells readily grow in serum-free medium; most, however, must be acclimated to a serum-free environment. This requirement is especially true if the cell line in question is already being cultured in serum-enriched medium (typically 10% v/v). It is possible to reduce serum content in medium; in some cases, it is advisable to do so. Reducing the amount of serum added can reduce cost, as serum is the most expensive component of the complete medium. Reducing serum also lowers the total protein content of the medium, facilitating collection of cell products, and minimizing sources of contamination. For cells growing in serum-enriched medium, a method of systematically reducing medium can be implemented. One must first consider the growth of cells in culture, before discussion of how to achieve serum reduction can begin (Figure 3.1). Cell growth in culture – whether the cells are adherent or suspended – is characterized by several stages. The lag phase, during which minimal or no cell division occurs, is a brief period after inoculation. The lag phase occurs as cells adjust to a new cell-culture environment, and as adherent cells begin the process of reattaching to the culture substrate. The lag phase is followed by the log or exponential phase. This is the major phase of cell division. The

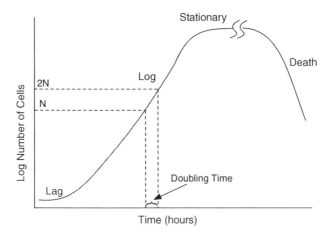

Figure 3.1   Cell growth in culture. The lag phase is indicative of minimal growth of a fresh sub-culture, followed by the log phase. Doubling time can be calculated during the log phase as the time it takes for the cell number to increase $2\times$. The stationary phase, where cell crowding and nutrient depletion slow cell proliferation, is followed by death of the population. For experiments where cell viability should be high and apoptosis rates low, samples should be acquired in the log phase or beginning of the stationary phase

doubling time, an indicator of cell growth, is determined during this period. The time for the cell population to double (Figure 3.1) can be determined at any point during the log phase, although it is most accurate at the center of that phase. After the log phase, the culture reaches the stationary phase. High cell density, contact inhibition, and consumption of nutrients signal a slowing of the cell cycle, and the cell concentration remains constant. Cell crowding, depletion of nutrients, and accumulation of waste eventually causes a sharp drop in cell concentration, called the death phase. This latter phase can be confirmed by microscopy, where the presence of a large number of dead cells, cell debris, and acidified medium (if an indicator is present) can be observed.

With an understanding of the growth phases of cultured cells, one can then design a method to reduce serum systematically. There are several methods of reducing serum, including switching to serum-free medium and observing cell growth. A more gradual approach is to culture the cells in the standard serum range (e.g., 10%), noting the doubling time during the log phase. For adherent cells, the doubling time can be estimated by counting the number of cells in the flask or plate, averaged over several fields of view. For suspended cells, a small aliquot of cell suspension can be counted on a hemacytometer (Chapter 2). On the next passage (Section 3.4), cells are diluted to the appropriate amount in medium with one-half of the serum

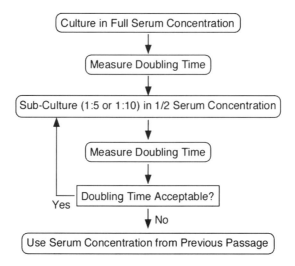

Figure 3.2 Schematic for a serum-reduction method. Sub-culturing in decreasing medium content, while monitoring the population doubling time (Figure 3.1) can be used to determine the minimum serum content

content (e.g., 5% in this example). The doubling time is again checked during the log phase for this second passage. The process is repeated, until the serum content is below approximately 0.1% (Figure 3.2). The next passage can then be serum-free. If, during any passage, the doubling time changes significantly, then the cells should be cultured in the serum content of the previous passage. In addition to doubling time, cell morphology should also be examined. If the cells have a particular function of interest (e.g., contractile behavior, insulin secretion, etc.), that function should be monitored along with the doubling time. Once either the doubling time or the function of interest diminishes, the serum reduction should stop. Once medium is diminished below 1–2% or eliminated altogether, some supplements may be required to maintain optimal growth. The most common additives are insulin, transferrin, and selenium. A combination insulin–transferrin–selenium supplement is available from many manufacturers to compliment serum-free medium.

The glucose content of basic medium varies (and is sometimes supplemented with additional glucose. The high glucose content of many medium types is intended to stimulate growth of the culture. However, some cell lines change phenotypic properties in high or low glucose [13,14]. When culturing for conditions close to those encountered *in vivo*, the glucose concentration should be adjusted to reflect the physiological value as much as possible. Like serum reduction, the impact of changes in glucose concentration can be monitored using the culture doubling time.

When formulating complete medium, care must be taken to preserve sterility of the final mixture. If all of the components are sterile to begin with, then aseptic handling in the biosafety cabinet will prevent contamination of the complete medium. If any of the reagents are not sterile at the onset, then filtration can be employed to remove contaminating organisms. When medium is prepared from powdered precursors (a less expensive, but more time-consuming approach), then filtration is necessary. Filter flask assemblies contain an upper chamber to hold the unsterile medium, a $0.22\,\mu m$ filter to prevent bacteria and fungi from entering the final product, and a medium bottle to collect the sterile medium. A vacuum pump is needed to drive the fluid into the lower chamber. The entire filter flask assembly comes in sterile packaging. This allows the medium to be formulated outside of the biosafety cabinet, and then transferred to the cabinet for sterile filtration.

3.3   THE USE OF MEDIUM IN ANALYSIS, AND ALTERNATIVES

Medium is primarily used to maintain cultures and samples before analysis. The medium can also be used during the analysis; in other instances, components of the medium may produce artifacts or otherwise interfere. As mentioned earlier in this chapter, and in Chapter 4, the presence of several components of medium can interfere with fluorescence measurements. Phenol red, one of the most common pH indicators added to medium, has a broad absorption band that interferes with most green fluorescence. Phenol red is also weakly fluorescent, creating an additional problem for green-emitting fluorophores. If the cell homeostasis is not required, then any buffer devoid of phenol red will work for fluorescence. If the cells are to be kept alive for long periods, then phenol-red-free medium is available from most medium manufacturers.

In addition to the weakly fluorescent properties of phenol red, other compounds present at relatively high concentration can interfere with fluorescence detection. Riboflavin is also weakly fluorescent, but the relatively large volume of the medium contributes to an unacceptable background signal. Proteins such as albumin, one of the major components of serum, also contribute strongly to autofluorescence of medium. For some epifluorescence microscopy, the background fluorescence from flavins and some proteins is minimal. For sensitive, single-molecule microscopy of cells, a buffer devoid of protein, indicator, and other additives must be used. Figure 3.3 shows a signal trace (intensity vs. time) of a 20 pM solution of rhodamine 110 in phosphate buffered saline and phosphate-buffered saline with 3% bovine serum albumin, both

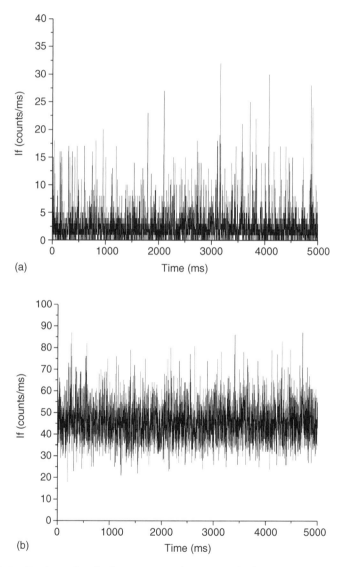

(a)

(b)

Figure 3.3 Single molecule fluorescence of 20 p.m. rhodamine 110 in phosphate buffered saline (a). When bovine serum albumin is present (b), the background fluorescence from the protein obscures faint fluorescence signals. For sensitive fluorescence measurements, medium that is both serum- and phenol-red-free should be used

excited by an argon-ion laser and detected on a single-molecule apparatus. The effect of the protein on the buffer is readily observed.

When cells, or cellular components, are separated by capillary or microfluidic electrophoresis (Chapter 5), the buffers typically used for separation are not compatible with culture. Physiological buffers, such as phosphate-buffered saline, typically contain approximately 135–150 mM sodium chloride to maintain osmolarity. Separation buffers often require a lower ionic concentration (typically 20–50 mM), and hence do not make good buffers to maintain cell function. It is possible to use additives to maintain osmolarity, while keeping the total ionic strength low. However, these buffers should not be used to store and separate cells or organelles. It is possible to use coated electrophoretic channels that are capable of handling buffers of physiological ionic strength. Using bilayer lipid coatings [15], glass and PDMS-based separation systems can be adapted for buffers that can maintain viability for longer periods of time than traditional separation buffers.

The exact medium used for culture depends on the cell type, the culture conditions, and the desired end result. For analysis, a similar selection process must be undertaken. The final medium or buffer used for analysis must be of low background, minimal interference, and – when possible – capable of sustaining cell viability and function for the experiment duration.

3.4 CULTURING CELLS

As cells grow in a culture flask or dish, they replicate, consume nutrients, and produce waste. As the culture progresses, these effects induce the stationary phase of cell growth. The process of sub-culturing cells requires some of the cells to be removed, and an appropriate volume of new medium added. For suspended cells, the process is straightforward. In a laminar flow biosafety cabinet (Chapter 2), a portion of the culture is aspirated from the flask or dish. At this point, it may be useful to count the cells on a hemacytometer (Chapter 2) to assess the cell concentration. The dilution ratio may then be determined. Alternatively, a routine dilution value can also be established. Less dilution (e.g., 1 : 5) will result in a higher sustained cell population, at the expense of more frequent sub-culture. A 1 : 10–1 : 5 ratio works well with most suspended cell types.

The culture of adherent cells is not significantly more difficult than suspended cells, but it is, at times, less convenient for analysis. Suspended cells can be maintained at high density in a flask at all times. Adherent cells must be removed from the flask by a protease, re-cultured, and allowed to settle, expand, and grow. This process means that there is an optimal time to harvest cells for analysis, and so timing of experiments is more important than when suspended cells are used. If other factors of experiment timing are more critical (e.g., the availability of a core instrument), then several, identical cultures can be maintained a few days apart from each other, so that at least one flask will always have the optimal cell density for the experiment. This process is more expensive and time consuming, but solves the issue of the cells being ready when the researcher is.

An example protocol for adherent cell sub-culture is given below. Variations on this theme can be made to accommodate cell types with special requirements. The procedure for suspended cell sub-culture is to first render the cells in the suspended state, then dilute them appropriately. A protease, such as trypsin, is added to the culture to dissolve the extracellular matrix keeping the cells on the surface. Figure 3.4 shows the process via microscopy. Before trypsin is added, the cells show the common phenotype of endothelial cells in culture. As time progresses, the

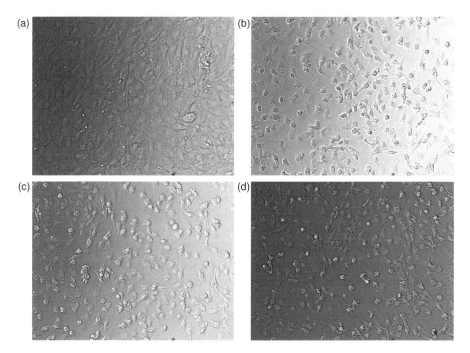

Figure 3.4   Mouse endothelial cells grown to high density in a culture flask (a) and after incubation with trypsin-EDTA solution at 1 min (b), 3 min (c), and 5 min (d). As the extracellular matrix is digested, cells detach and enter suspension for harvesting and sub-culture

cells retract into spherical shape as they leave the surface. Prolonged exposure to proteases such as trypsin can result in loss of cell surface antigens, and ultimately cell death. Therefore, cells should be kept in trypsin solution until all of the cells detached, then the reaction is quenched (see the following protocol). Cells can also be scraped from the surface, but this method is more disruptive than protease release. Since, in many cases, cells must be either rendered into suspension, or transferred to another vessel for analysis, the sub-culture event can coincide with analysis to maximize both reagent use and time.

PROTOCOL 3.1: SUB-CULTURE OF ADHERENT CELLS

Cells should be confluent or nearly so (Figure 3.4). Warm all reagents to 37 °C in a water bath and sterilize the culture hood before retrieving the culture from the incubator. All fluid handling is performed in a biosafety cabinet.

1. Using an autopipetter, aspirate the cell-culture medium (Figure 3.5) as much as possible. Discard the serological pipette (SP) containing the medium.
2. Using a new SP, rinse the surface containing the cells gently with 5–10 ml sterile phosphate-buffered saline. Aspirate the phosphate-buffered saline with the same SP, keeping the pipette in the flask the entire time. Note: avoid scraping the flask bottom with the SP tip.
3. Introduce 5–8 ml of trypsin–EDTA solution (0.25% Trypsin, 0.02–0.1% EDTA in Hank's Balanced Salt solution) using a new SP. Place culture flask in incubator for 3–5 minutes, or until all cells detach from surface (see Figure 3.4 for example images).
4. Once cells have detached, add 10–12 ml of culture medium to the flask to quench the protease reaction.
5. Aspirate 80–90% of the solution (depending on dilution ratio). Note: This aspirated solution can be used directly for analysis if suspended cells are required, or cells can be transferred to another culture device and allowed to reattach for analysis. This cell suspension can also be centrifuged and cells resuspended in medium or a buffer, if needed.
6. Replace an equivalent volume of fresh culture medium into the original culture flask with a new SP.

New flasks should be used (i.e., the remaining solution from step 5 should be transferred to a new flask before proceeding) every few sub-cultures. When sub-culturing cells for microscopy, they can be seeded into

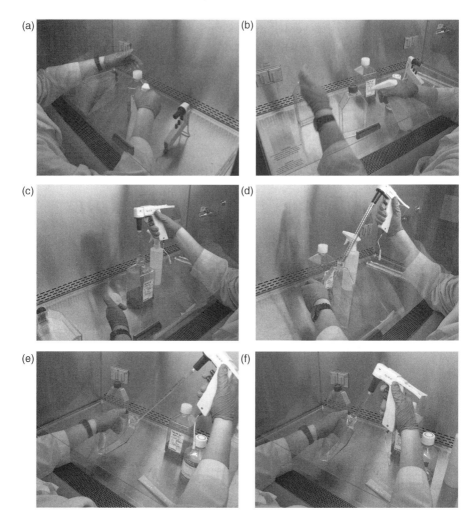

Figure 3.5   Fluid handling in a biosafety cabinet. Work should be performed in the center of the workspace. Bottle necks should be sprayed with 70% ethanol before opening (a and b). Serological pipettes should be inserted into medium bottles (c) so that the tip does not touch the sides of the bottle neck. The same pipette (containing medium) should then be transferred to the culture flask without touching the tip to the neck sides before dispensing the fluid (d). When washing adherent cells before incubation with trypsin, gently dispense warmed saline while moving the tip cross the surface of the flask for even surface coverage (e). When aspirating liquids or cell suspensions, angle the flask to remove as much fluid as possible (f)

a variety of culture dishes, chambers, and plates designed for fluorescence or white-light imaging (see Chapter 4).

3.5   GROWING CELLS IN THREE DIMENSIONS

Cell growth *in vivo* is a three-dimensional (3D) process, complete with supporting cells, layered cell types, and a complex environment that is difficult to reproduce *in vitro*. Since most cell cultures are two-dimensional (2D) flasks, plates, or similar systems, there is a fundamental difference between cellular environments. The presence of extracellular matrix, and the interaction of cell contact in three dimensions, is not reproduced in 2D culture. For adherent cells, it is possible to grow cells in 3D cultures that can approach *in vivo* phenotype and functionality. It is still difficult, in these cases, to reproduce the exact conditions of the intact organism, but in many cases these 3D tissue models offer remarkable advantages to 2D culture.

Solid support and fluid-transport mechanisms are required in order to grow cells in 3D culture (Figure 3.6). There are several methods to accomplish this, including stirred flasks, roller bottles, and rotating-wall bioreactors that all produce cultures that differ from regular 2D methods. Stirred flasks can be operated in batch-fed (periodic medium replacement) or perfused (continuous medium flow) modes. Most stirred flasks have an agitator or stirrer in the flask during cell growth. The high-shear environment can result in excessive cell damage, and impact

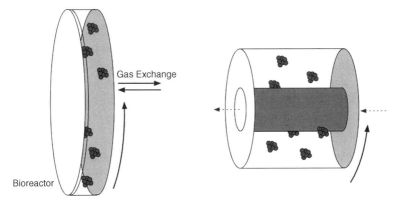

Figure 3.6   High aspect-ratio (a) and perfused, filter-core (b) rotating-wall bioreactors. In these designs, constant rotation of both walls produces low shear for culturing cells in three dimensions. In (a), a large membrane on the back wall allows for gas exchange; feeding is done by batch replacement of medium. In (b), a continuous medium flow through the filtered core is possible, allowing for culture medium to be replenished continuously

between 3D bodies and the stirring element can also impede tissue formation. Roller bottles also suspend cells and scaffold or support materials, allowing aggregation and 3D growth. However, the gradient of fluid shear between the bottle wall and the center adversely affect culture conditions. Rotating-wall bioreactors feature either an inner and outer core or two parallel walls with a high aspect ratio. This approach minimizes shear and allows tissues to form that cannot be generated by other means. Rotating-wall bioreactors are available commercially from Synthecon and other suppliers, or can be made in machine shops that are familiar machining with Teflon, delrin, and other polymeric materials.

Rotating-wall bioreactors have been used to generate a variety of tissues [16–18], using scaffolds in suspension in the rotating vessel. When growing cultures in rotating vessels, culture bodies of significant size (>1 mm) can be obtained after a few days of culture. When growing cells in this manner, care must be taken to ensure efficient mass transport into and out of the cell mass. Cell layers only a few hundred micrometers thick show decreased mass transport and a gradient of nutrients and waste throughout the tissue. Scaffolds that are porous or contain a channeled structure can be used to improve mass transport, but in increasing tissue masses, the likelihood that the core consists of dead cells is greater. Confocal microscopy can be used to assess the viability and function of 3D culture bodies, although two-photon microscopy will offer better penetration depth and more accurate measurements (Chapter 4). A tracer compound can also be incubated with the culture, and then fluorescence measurements used to assess penetration of the tracer into the tissue mass.

Three-dimensional cultures can also be made using embedded gels, allowing a variety of culture types to be grown without the need for bioreactors [19]. Growing cells in an embedded-gel approach allows culture sizes to be confined to smaller sizes amenable to confocal imaging. The culture bodies can be left in the gel, or released and placed in other arrangements and arrays (Figure 3.7). Since the cell number is limited, many cell-based assays can be probed by microscopy without worry of mass transport to internal cells.

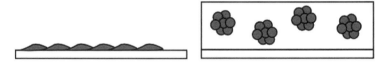

Figure 3.7   Three-dimensional microcultures using an embedded-hydrogel technique. In contrast to 2D culture (a), small numbers of cells can be grown in three-dimensional bodies using a hydrogel to confine the cells (b). The hydrogel can be removed, allowing the culture aggregates to be analyzed by various methods (adapted from [19])

3.6   STERILITY AND CONTAMINATION OF CULTURE

As discussed in Chapter 2, cell culture is a constant struggle to keep opportunistic organisms out of the culture environment. Isolation techniques, such as biosafety cabinets, and proper culture equipment are the best defense against contamination. Carefully followed lab protocols, and the avoidance of working with two different culture flasks at once, will also minimize cross-contamination between cell lines. It is also possible to add antibiotic and antifungal supplements to culture medium to reduce the risk of bacterial and fungal contamination. When isolating primary cells, it is necessary to add these reagents to prevent the spread of organisms from the donor tissue. However, in many cases the subsequent cultures are maintained in a low concentration of antimicrobial/antifungal reagents such as penicillin and streptomycin. Immortalized cell lines are also often maintained with antibiotics. The debate between using and omitting these reagents is that culture medium, ideally, should mimic *in vivo* conditions as closely as possible. The addition of antimicrobial/antifungal reagents can, in some cases, limit or alter some cell functions. Cells should be grown with and without antimicrobial/antifungal reagents and the desired cell function assessed in both conditions.

Another, more universal concern of using these reagents is that they only prevent *some* organisms from entering the culture. Low-level contaminations of antibiotic-resistant bacteria or antifungal-resistant fungi can go undetected in these cases. Since the addition of these reagents creates a false sense of security in the sterility of the culture, these types of low-level contaminations may alter cell function as well. A low concentration of resistant bacteria can evade detection indefinitely, and create artifacts in a variety of genomic, proteomic, or whole-cell assays. In addition, the use of antimicrobial and antifungal reagents does not absolve researchers from exercising care when handling cultures, medium, and other sterile materials. It is conceivable that, if one is careful, these reagents are not needed. Despite the controversy when using them, they do reduce the risk of a major contamination, which can set back research for several weeks (see Chapter 2 for steps on isolating a persistent contamination).

While long-term use of antibiotics is not recommended in many cases, it is possible to treat infected samples with a temporary antibiotic regimen to reduce or eliminate infections, particularly for mycoplasma [20]. The use of antibiotic treatment, combined with altered culture conditions and daily medium exchange, may rid a cell line of mycoplasma. The regimen includes culturing daily in higher serum content medium, adding anti-biotics directly to the culture, and growing the cells at higher density. The authors of this approach also note that apoptosis of the cell line is possible due to accumulated cytotoxic effects of some antibiotics. It is possible, therefore, to include an apoptosis inhibitor to prevent culture loss during antibiotic treatment.

3.7   STORAGE OF CELL SAMPLES AND CELL LINES

Over years of cell research, laboratories can acquire tens or hundreds of cell lines, particularly if they are derived from primary donors. In many cases, it is impractical to keep every culture in the lab in active growth and medium exchange. A cell line used once a year will require continual sub-culture to maintain viability. It is therefore better to store cells in a manner that requires no consumption of reagents. The solution, refined through years of animal husbandry, is to cryopreserve (freeze) cells. Cryopreserva-tion, when performed properly, places the cells in a type of suspended animation, where all metabolism and mass transport stops and the cells remain in the same state over time. Of course, the act of cryopreservation can alter cells, although careful optimization of freezing and thawing conditions can minimize these effects.

The basic principle of cryopreservation is to freeze the cells in a controlled way, ideally so vitrification occurs. Vitrification, or conversion to a glass-like state, maintains viability. Retrieval of cells from cryogenic storage requires controlled return to an aqueous state. Cryopreservation is therefore a struggle to maintain cell viability and function during pre-servation and also upon thawing. To fail in either regard is to subject the cells to serious damage. To understand how this process developed, it is useful to consider the earliest work in the field. Cryopreservation has been used for decades in animal breeding, with animal semen being the primary sample for storage. When one considers the unique nature of semen for reproductive purposes, the high cell concentration, and the ability to concentrate or vary the amount of sample delivered to the female animal, means that large losses of viable cells do not hinder the goal of reproduc-tion. In plain terms, a loss of 30% of the cells means that there are still adequate numbers of sperm cells to complete fertilization. When preser-ving more precious cells, however, cryopreservation must be carefully controlled. Consider female gametes, which are typically in lower supply. A 30% loss of egg cells would be unacceptable in many cases. As cryopreservation became more widely applied, the methods of storing sperm cells had to be modified to accommodate other cells, including using post-storage reagents to preserve cell viability.

To understand why cells die during cryopreservation, it is important to discuss what happens during cryopreservation (Figure 3.8). As the temperature decreases in the sample, extracellular ice crystals begin to form. This process can affect cells in two ways. First, as the crystals grow too rapidly, they may physically disrupt the cells. Second, as more water enters the ice crystal, the concentration of dissolved materials outside of the cell (salts, etc.) increases. Inside the cell, smaller ice crystals can also form, which can disrupt the internal structure of the cell. When freezing, the rate of temperature decrease must be carefully balanced. If the sample is frozen too quickly, extracellular and intracellular ice formation can destroy the cells. If the sample is frozen too slowly, cells can die from exposure to the increasingly hypertonic solution. The same process can occur during thawing.

Given the delicate balance that must be achieved – and optimized for each cell line – cryopreservation can be a delicate task. There are, fortunately, additives that can improve chances of cell survival during storage and recovery. Additives can minimize ice formation inside or outside the cell, or impede cell death. In an ideal case, no ice crystals would form, and the cells would transition from liquid to a vitrified (glassy) state. This vitrification point must be carefully monitored, as at any point ice formation can occur. Many different sugars, proteins, and other additives have been reported as cryopreservation agents. Two of the most common methods – found in many proprietary cryopreservation solutions – are dimethylsulfoxide (DMSO) and glycerol. DMSO serves to lower the freezing point of cells and is a useful cryopreservative because it easily permeates cells and tissue samples. It is, however, cytotoxic at higher concentrations and prolonged exposures. A 5–10% solution of DMSO in culture medium works well as a cryopreservation medium. Serum can be added to 10% v/v ratio to this medium (i.e., 10% serum, 5–10% DMSO, v/v). The DMSO concentration should be as low as possible to reduce toxicity, without resulting in high loss of viability during cryopreservation. Glycerol also reduces ice formation and cell dehydration, and is less cytotoxic than DMSO. However, glycerol does not have the same viscosity and cell permeability. For cell clusters, tissues, or multi-cellular structures, DMSO may be a better alternative than glycerol. In both cases, sterile, high-quality reagents should be used to avoid contamination of the culture.

The rate of freezing and thawing also plays an important role in minimizing intracellular and extracellular ice formation. For mammalian cells, a freezing rate of approximately 1 °C per minute will achieve

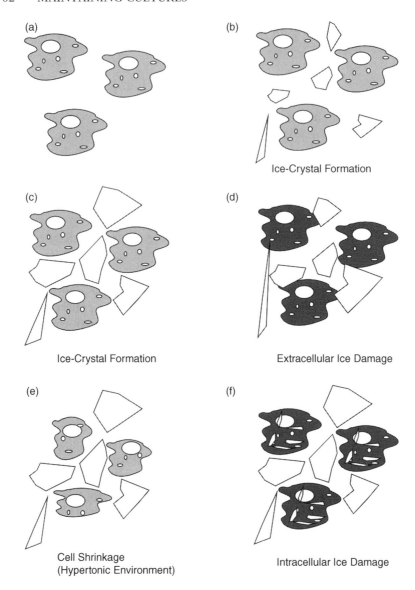

(a)

(b)

Ice-Crystal Formation

(c)

Ice-Crystal Formation

(d)

Extracellular Ice Damage

(e)

Cell Shrinkage
(Hypertonic Environment)

(f)

Intracellular Ice Damage

Figure 3.8   During cryopreservation of cells (a), ice crystals form in the extracellular fluid (b). Careful control of ice-crystal formation can yield high recovery upon thawing. As ice crystals enlarge (c), they can mechanically damage cells, resulting in loss of viability (d, dead cells shown as dark gray). During ice-crystal growth, the increasingly hypertonic solution can result in cell dehydration and death (e). As the cells cool or thaw, intracellular ice crystals may also form, mechanically damaging the cells (f). The same processes can occur during cell thawing. Controlled freezing/ thawing, and the use of cryopreservatives and apoptosis inhibitors can increase the yield of cryopreserved cells acceptable viability. Several manufactures sell containers designed to promote freezing near the 1 °C per minute rate. These containers contain isopropanol or other materials designed to regulate the inner chamber. The protocol listed in this chapter does not require one of these devices; however, for sensitive or rare cells, one may be required.

PROTOCOL 3.2: CRYOPRESERVATION OF MAMMALIAN CELLS

Note: It is best to start with a healthy population in the log phase, with high cell density. Check cells via microscopy and fluorescence microscopy for contamination (Chapters 1 and 2). Only work with one cell line at a time to prevent cross-contamination.

With the exception of centrifugation, all work is conducted in a laminar biosafety cabinet.

1. For adherent cells, place into suspension (See Protocol 3.1 on adherent cell sub-culture) using trypsin or a similar protease. The quenched solution (released cells + trypsin + medium w/10% serum) will be used for cryopreservation.
2. For suspended cells, an aliquot of the culture flask will be used.
3. Using a serological pipette, transfer the cell suspension to a sterile 50 ml centrifuge tube (or 15 ml, if the aliquot is smaller).
4. Centrifuge at $1000 \times g$ for 4 minutes to produce a pellet of cells. Discard supernatant.
5. Re-suspend cells in medium supplemented with 5–10% DMSO (note: 5% DMSO is recommended for most cases) or glycerol. Note: The amount of cryopreservation solution used depends on the original culture sample and how many 1 ml cryopreservation vials will be filled. For example, a 25 ml culture suspension can be centrifuged and resuspended in 10–12 ml of cryopreservation solution, filling 12–15 vials.
6. Aliquot just less than 1 ml of suspension into cryovials. Each vial should then be capped tightly and labeled with the cell line, the date, and the individual performing the cryopreservation. Other pertinent notes can be added to the log book.
7. Place cryovials into −70--80 °C freezer for at least 3–6 hours, or overnight. Note: Place in cryopreservation container if more controlled cooling is needed.
8. Transfer cryovials to liquid nitrogen vapor phase.

A log book is particularly useful for liquid nitrogen storage, as the number of vials stored in racks can impede fast retrieval. The log book should also contain other information that cannot be written on the vial. The passage number, cell source (if primary, the animal or patient number should also be noted), and any other information should be recorded.

When retrieving cells from liquid-nitrogen storage, a similar level of care is required to prevent the same processes outlined in Figure 3.8. A controlled thawing rate is required to maximize cell recovery. Cell retrieval is a more critical step, since the chance for contamination is higher, and there is an added safety element as well. As mentioned earlier, cryovials may contain a small amount of liquid nitrogen (this is especially true if the cryovial was accidentally submerged in the liquid phase of the Dewar). The expansion ratio of liquid nitrogen is sufficient to cause the vial to explode shortly after the vial is exposed to room temperature. A face shield, goggles, and personal protective equipment should be used during the thawing phase (Chapter 2). Alternatively, the vial can be placed in the biosafety cabinet and quickly "vented" (i.e., the cap loosened and then tightened again) to remove any nitrogen gas. The vial is then ready for thawing. Of the many commercial heating systems designed for controlled warming, placing the vial partly into a water bath (Protocol 3.3) is perhaps the easiest method to thaw the sample. The cell sample can be used upon thawing, or centrifuged to remove the cryopreservation solution. If higher concentrations of DMSO are used, then this step can reduce cytotoxicity effects. If the sample is not centrifuged, but the suspension (in cryopreservative) is transferred to the culture vessel, then it should be monitored to ensure that the diluted preservation agent does not affect cell function. The thawed sample is then transferred to a biosafety cabinet for use or the incubator for culture. The following protocol is a general one that works for many mammalian cell types. As always, for precious cells more optimization will be required.

PROTOCOL 3.3: RETRIEVAL OF CELLS FROM LIQUID-NITROGEN STORAGE

Note: Liquid-nitrogen expansion may cause the vial to rupture. Vent the vial when possible, and wear appropriate safety equipment.

1. Fill a culture vessel (flask, dish, etc.) with sterile medium warmed to 37 °C, place in incubator.
2. Retrieve the cryovial from liquid nitrogen storage. Note: Wear cold-resistant gloves over lab gloves to avoid frostbite.

3. Place cryovial in biosafety cabinet and briefly vent (<1 second). Tighten vial and transfer to a 37 °C water bath.

4. Holding the vial with the cap and cap threads *above* the water line, gently swirl the vial in the water bath. It is important to make sure the vial does not submerge, as it can become contaminated. Note: Check periodically to determine when the solution in the vial has melted completely.

5. If centrifugation of the cell suspension is required, centrifuge at the appropriate speed and time before proceeding to step 6.

6. After the solution has melted, transfer the vial to the biosafety cabinet. Retrieve the flask or dish containing medium from the incubator.

7. Spray the culture flask cap and neck (or outside of dish) with 70% ethanol to sterilize. Spray the outside of the cryovial.

8. If the sample was centrifuged (step 5), aspirate the cryopreservation solution and replace with an equivalent volume of warmed culture medium.

9. Using a serological pipette, transfer the cell suspension to the culture vessel and place in the incubator.

10. Monitor the culture for 1–3 days to ensure proper cell growth and morphology.

Upon thawing and subsequent culture, a percentage of the cells will have died from mechanical damage (ice and dehydration). Another portion of the cell population will die after some time from apoptosis [21]. The stress of freezing and recovery can trigger this secondary death mechanism. It is possible to reduce the effects of apoptosis loss by adding an apoptosis inhibitor, such as the broad caspase inhibitor Z-VAD-FMK [22] to the culture. Adding an inhibitor during the freezing, thawing, and initial, subsequent culture step can improve cell viability. When validating a cryopreservation protocol for the first time with a particular cell line, it is useful to test the cells before and after cryopreservation for both morphology and apoptosis. The latter should be checked a few hours after thawing and the following day as well. Chapter 9 lists several protocols for apoptosis monitoring (see Chapters 4 and 6 for background on microscopy and flow cytometry). As mentioned in Chapter 2, frozen cell stocks can be stored in the vapor phase of a liquid nitrogen Dewar, or they can be stored in a −80 °C freezer. The latter approach has a risk of losing the entire collection of cell stocks if power is lost or the refrigerator malfunctions. With liquid nitrogen storage, care must be taken to avoid submerging the cells in the liquid phase.

3.8   CONCLUSION

Cell analysis requires a laboratory capable of maintaining healthy cell samples from a variety of sources. The primary component of cell maintenance, medium, varies in composition and complexity, based on the needs of the cell line. While it is important to select the best medium for a given cell line, many cells can be grown well in a small handful of medium formulations (Table 3.1). The added components of complete medium also affect, not only cell growth, but the analysis as well. Serum, while necessary for many cell lines, complicates many proteomic and spectroscopic analysis. Serum-free medium should be used for culture in these cases, even for a brief period to rid the cells of residual serum before analysis. Other additives, such as antibiotics, should be evaluated before use to determine if they are necessary. Antibiotics and antifungals only rid the culture of virulent, visible contamination. Low-grade contamination of resistant bacteria or fungi may persist indefinitely.

Cryopreservation of cell lines is critical to any cell-analysis laboratory. It allows for archiving of cell samples and storage of backup stocks in the event of incubator failure or contamination. Efficient use of cryopreservation reduces wasting of reagents to maintain a culture that will not be used for months or longer. The cryopreservation protocols listed in this chapter are a guide that work well with many cell lines, but may have to be refined for more sensitive cells. Inhibition of apoptosis, and the use of cryopreservatives, may improve cell recovery in the cryopreservation process.

Regardless of the type of cell analysis discussed in subsequent chapters, the conditions of culture can affect the outcome of the experiment. For example, if apoptosis induction agents are to be studied, the background level of apoptosis in the cell population can vary depending on how fresh the sub-culture is. By maintaining a careful regimen of cell feeding before analysis, a less variable apoptotic population will be obtained, allowing meaningful results to be generated. Careful documentation of sub-culture, efficient use of reagents, and sterile practices will lead to a cell analysis laboratory that is capable of realizing the full potential of any analytical technique.

REFERENCES

1. Wang, K., Cometti, B., and Pappas, D. (2007) Isolation and counting of multiple cell types using an affinity separation device. *Analytica Chimica Acta*, **601**, 1–9.

2. Liu, K., Dang, D., Harrington, T. *et al.* (2008) Cell culture chip with low-shear mass transport. *Langmuir*, **24**, 5955–5960.

3. Wang, K., Marshall, M.K., Garza, G., and Pappas, D. (2008) open-tubular capillary cell affinity chromatography: single and tandem blood cell separation. *Analytical Chemistry*, **80**, 2118–2124.

4. Wang, K., Solis-Wever, X., Aguas, C. *et al.* (2009) differential mobility cytometry. *Analytical Chemistry*, **81**, 3334–3343.

5. Wheeler, A.R., Throndset, W.R., Whelan, R.J. *et al.* (2003) Microfluidic device for single cell analysis. *Analytical Chemistry*, **75**, 3249–3254.

6. Rettig, J.R. and Folch, A. (2005) Large-scale single-cell trapping and imaging using microwell arrays. *Analytical Chemistry*, **77**, 5628–5634.

7. Hu, S., Krylov, S., and Dovichi, N.J. (2003) Cell cycle-dependent protein fingerprint from a single cancer cell: image cytometry coupled with single-cell capillary sieving electrophoresis. *Analytical Chemistry*, **75**, 3495–3501.

8. Sims, C.E., Meredith, G.D., Krasieva, T.B. *et al.* (1998) Laser-micropipet combination for single-cell analysis. *Analytical Chemistry*, **70**, 4570–4577.

9. Claycomb, W.C., Lanson, N.A., Stallworth, B.S. *et al.* (1998) HL-1 cells: a cardiac muscle cell line that contracts and retains phenotypic characteristics of the adult cardiomyocyte. *Proceedings of the National Academy of Sciences USA*, **95**, 2979–2984.

10. Benincasa, M.-A., Moore, L.R., Williams, S. *et al.* (2005) Cell sorting by one gravity SPLITT fractionation. *Analytical Chemistry*, **77**, 5294–5301.

11. Rubakhim, S.S., Page, J.S., Monroe, B.R., and Sweedler, J.V. (2001) Analysis of cellular release using capillary electrophoresis and matrix-assisted laser desorption/ionization-time of flight mass spectrometry. *Electrophoresis*, **22**, 3752–3758.

12. Reisser, D., Fady, C., Pelletier, H. *et al.* (1989) Comparative effect of rat and fetal calf serum on measurement of the natural tumoricidal activity of rat lymphocytes, macrophages and polymorphonuclear cells. *Cancer Immunology and Immunotherapy*, **28**, 34–36.

13. Cechowska-Pasko, M., Palka, J., and Bankowski, E. (2007) Glucose-depleted medium reduces the collagen content of human skin fibroblast cultures. *Molecular and Cellular Biochemistry*, **305**, 79–85.

14. Gstraunthaler, G., Seppi, T., and Pfaller, W. (1999) Impact of culture conditions, culture media volumes, and glucose content on metabolic properties of renal epithelial cell cultures. *Cellular Physiology and Biochemistry*, **9**, 150–172.

15. Phillips, K.S., Kottegoda, S., Kang, K.M. *et al.* (2008) Separations in poly(dimethylsiloxane) microchips coated with supported bilayer membranes. *Analytical Chemistry*, **80**, 9756–9762.

16. Hwang, Y.S., Cho, J., Tay, F. *et al.* (2008) The use of murine embryonic stem cells, alginate encapsulation, and rotary microgravity bioreactor in bone tissue engineering. *Biomaterials*, **30**, 499–507.

17. Rivera-Solorio, I. and Kleis, S.J. (2006) Model of the mass transport to the surface of animal cells cultured in a rotating bioreactor operated in micro gravity. *Biotechnology and Bioengineering*, **94**, 495–504.

18. Bursac, N., Papadaki, M., White, J.A. *et al.* (2003) Cultivation in rotating bioreactors promotes maintenance of cardiac myocyte electrophysiology and molecular properties. *Tissue Engineering*, **9**, 1243–1253.

19. Lee, G.Y., Kenny, P.A., Lee, E.H., and Bissell, M.J. (2007) Three-dimensional culture models of normal and malignant breast epithelial cells. *Nature Methods*, **4**, 359–365.
20. Uphoff, C.C., Meyer, C., and Drexler, H.G. (2002) Elimination of mycoplasma from leukemia lymphoma cell lines using antibiotics. *Leukemia*, **16**, 284–288.
21. Men, H., Monson, R.L., Parrish, J.J., and Rutledge, J.J. (2003) Degeneration of cryopreserved bovine oocytes via apoptosis during subsequent culture. *Cryobiology*, **47**, 73–81.
22. Heng, B.C., Clement, M.V., and Cao, T. (2007) Caspase inhibitor Z-VAD-FMK enhances the freeze-thaw survival rate of human embryonic stem cells. *Bioscience Reports*, **27**, 257–264.

4

Microscopy of Cells

4.1 INTRODUCTION

Microscopes and cells have enjoyed a 340-plus year together, and for good reason. The length scale of the typical cell – be it bacterial, mammalian, or plant – matches the resolution obtained from the most basic optical microscopes. It is therefore little surprise that even toy microscopes – now equipped with digital cameras and a direct interface to a computer – can be used to image cells. Hooke's earliest experiments unearthed a hidden world of cell structure and function; experiments in the nineteenth century revealed many organelles. Since those discoveries, microscopes have evolved into complex and high-resolution spectroscopy instruments capable of the spatial, spectral, and temporal resolution needed to study cells.

For all of the benefits of cell microscopy, there are several pitfalls that can quickly ruin even the best-laid experiment. Unfortunately, many of the techniques of cell microscopy almost require an apprenticeship to learn. Unwritten secrets, small variations of the theme that rarely appear in research publications, are critical for successful experimentation. There are also seemingly as many commercial cell accessories for microscopy as there are cell types to study. Attachments to microscopes include incubation systems, manipulation stages, and so on. These accessories fall into two categories: essential (i.e., necessary to conduct an experiment) and useful (they make life easier, but are not critical). This book will only deal with the former.

Practical Cell Analysis   Dimitri Pappas
© 2010 John Wiley & Sons, Ltd

There is a wide palette of options when cells are observed on a typical research or clinical microscope. The instrumentation itself varies according to both need and budget. For example, routine analysis of sperm cells for reproductive medicine requires a simple transmission microscope, possibly equipped with a digital camera and counting software. On the other end of the spectrum, measuring calcium signaling in cells can require two-photon excitation and a laser-scanning confocal microscope. The difference in complexity – and cost – can be staggering. It is therefore of primary importance that the types of cellular analysis are known (or at the very least anticipated) prior to acquiring an instrument, or seeking out core facilities.

As in any analytical or clinical setting, one must consider the tasks to be completed in both routine operations and the occasional, out of the ordinary test. Chapter 9 of this book compiles cell analyses, protocols, and probes, many of which are amenable to microscopic investigation. However, one must also consider not only the *type* of analysis, but also the figures of merit one hopes to achieve. For example, it is not enough to decide that cell viability is important. Measuring cell fluorescence by propidium iodide – or other membrane impermeable dye – will identify dead cells. Is the end user interested in the shape of the nucleus? Is the degree of chromatin fragmentation important? Is temporal resolution important? One must also consider the statistics involved in measuring cells (see Chapter 8 for a detailed discussion of cell counting and statistics). In most cases, a compromise between figures of merit must be reached. The need to count many cells imposes limitations of the microscopy field of view (and therefore resolution). In confocal scanning methods, the need to observe the 3D volume of a cell and surrounding space also restricts temporal resolution. As in most analyses, therefore, one must choose the optimum, yet balanced, solution before proceeding with what might be a rare or expensive sample.

4.2   MICROSCOPE TYPES

This chapter discusses the various microscopy methods commonly used to analyze cells. In addition to the optical microscopy methods briefly discussed in the introduction (Section 4.1), nonoptical methods are also discussed. Both atomic force microscopy (AFM) and scanning electron microscopy (SEM) are routinely applied to cellular analysis. In the case of the latter, so-called environmental scanning electron microscopy (ESEM) provides cell imaging without complete desiccation of the sample.

However, it should be noted that the cell would not survive any SEM treatment.

As mentioned above, many microscopy techniques can be applied to cellular analysis. Fortunately, many universities, hospitals, and institutes operate core facilities where many different microscopy instruments are located in a single laboratory. The limitations of budget, space, and upkeep can therefore be reduced for individual investigators. Whether acquiring an instrument for one's own laboratory, or choosing which core instruments to use, it is important to consider both the sample and the desired end result(s). Figure 4.1 outlines the necessary parameters and figures of merit one might consider when choosing the appropriate microscopy method.

The simplest microscope type is the transmission microscope (Figure 4.2). Transmission microscopes are found in most lab settings

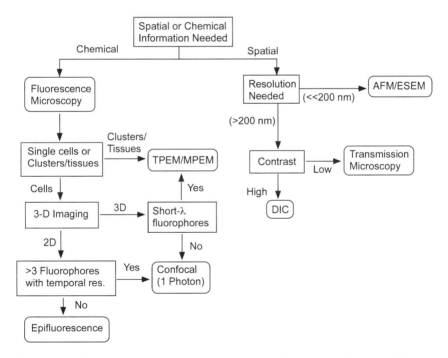

Figure 4.1 Flow chart for selecting the appropriate microscopy technique. Cell type, size, and morphology must be considered, along with the degree of spatial resolution and chemical information desired. AFM = Atomic Force Microscopy; ESEM = Environmental Scanning Electron Microscopy; DIC = Differential Image Contrast; TPEM = Two-Photon Excitation Microscopy; MPEM = Multi-Photon Excitation Microscopy

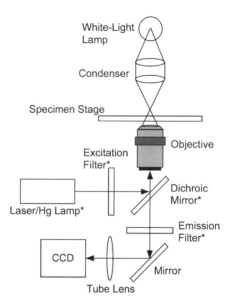

Figure 4.2 The basic components of transmission (white-light) and epifluorescence microscopes. Components with an asterisk(*) are required only for fluorescence systems

and can be acquired for minimal investment. For light-transmission microscopes, fixed-tube length or infinity systems offer similar performance (the former can be assembled in a laboratory for less than $500). More modern systems feature polarization, differential image contrast (DIC), and brightfield/darkfield options. Both DIC and darkfield offer improved contrast, although the latter suffers with regard to spatial resolution. In addition to the standard tungsten-type white-light sources, compact microscopes using LEDs are also available. An added benefit of the LED-based systems is that many are portable, allowing for in-field measurements of cells.

Most optical microscope objectives work with air between the front surface of the outermost lens and the specimen or coverslip. These so-called "dry" objectives feature numerical apertures between 0.1–0.7 for most conditions. For higher numerical aperture (and therefore higher resolution), an immersion medium is used. For oil immersion, the microscope can either be upright (the traditional microscope style) or inverted. Inverted microscopes are generally more amenable to biological specimens, especially when the objective working distance is short (as it often is with oil-immersion objectives). Water-immersion objectives also feature high numerical aperture. In this case, the objective can function with a coverslip over the specimen and a droplet of water coupling the objective to the glass, as in oil immersion. Water immersion objectives can also be dipped directly into the specimen (up to a certain depth). While this approach is tempting due to its simplicity, it should generally be avoided unless micromanipulation of the cell is also necessary. The contamination issues of directly inserting the objective into a specimen are great; considering the high cost of these objectives, this approach should be used with caution. Another drawback of the water-immersion objective is that it is difficult to use with an inverted microscope. The viscosity of water is much lower than most immersion oils, and therefore it is difficult to maintain a coupling droplet in an inverted geometry. Evaporation of the water coupling droplet must also be considered, and is generally not an issue for oil systems.

The spatial resolution of any microscope is determined largely by the numerical aperture, magnification, the field of view, and the size/resolution of the readout device in the image plane. The field of view (i.e., number of cells) will decrease as the spatial resolution improves, so the best and necessary magnification must be chosen. When counting leukocytes in a blood smear using a fluorescent dye, such as one of the Hoechst DNA dyes, total magnifications of 40–200× are typically sufficient. For example, counting blood cells on a hemacytometer mounted on an Olympus IX71 inverted microscope requires a total magnification of 100× to observe red blood cells or 40× to observe white blood cells.

Light-transmission microscopy is the most straightforward – and oldest – technique to study cells. Contrast, however, is poor. Mounting a cell specimen often requires a colored stain that (ideally) stains only one type of cell, or allows visual differentiation. One of the most routine analyses in this regard is the Pap smear, which ideally allows objective identification of abnormal cervical cells. Optical methods of producing contrast, such as differential image contrast, create a "shadowed" effect, without the use of staining dyes, which is useful for cell morphology measurements. It is important to note that fixed cells or cells that have been permeabilized typically display poor contrast characteristics.

The colored stains used in transmission microscopy offer some biological and chemical selectivity. However, fluorescence microscopy typically offers greater selectivity with significantly improved contrast. At the highest end of performance, single molecules can be detected and studied within cells [1–3]. Fluorescence microscopes are typically modified from the transmission microscope base (Figure 4.2). The most common – and accessible – type is the epifluorescence microscope. Illumination is roughly uniform across the field of view (Note: this is not guaranteed, and can be quantified using a thin film of dye between two coverslips). Fluorescence sources are brighter than white-light sources, which can cause problems when imaging live cells (see phototoxicity, below).

Epifluorescence, while simple to use and sensitive, suffers from several artifacts. Out-of-focus light, which is typically less of an issue in transmission microscopy, can generate offset signals that make quantitative measurements of fluorescence difficult. The confocal microscope, however, eliminates these and other artifacts by placing a pinhole in the conjugate image plane of the microscope. The confocal microscope, while eliminating out-of-focus light and rejecting more background light, also provides an increase in spatial resolution. Confocal instruments can either be laser scanning or Nipkow disk systems. Laser-scanning systems have higher excitation irradiance and multi-color capability, while Nipkow disk systems are typically faster from an acquisition standpoint.

The question could then be raised, why not always use confocal microscopes? Generally speaking, laser-scanning confocal microscopes are expensive, although they can be built in the lab in a straightforward approach [4,5]. Confocal instruments are also less robust and require a higher expertise to operate than a typical epifluorescent microscope. However, confocal images yield resolution and sensitivity that are sometimes not only desired, but critical to an analysis. Resolution along the optical axis (z-axis) requires confocal detection. In addition to intensity measurements, more confocal instruments (commercial and home-built) are now adding additional fluorescence measurements to the palette of options. For example, fluorescence correlation and cross-correlation spectroscopy (FCS and FCCS), fluorescence lifetime imaging microscopy (FLIM), and fluorescence recovery after photobleaching (FRAP), are some of the many techniques that require minimal or no modification to an existing confocal microscope. FCS delivers information on intracellular analyte concentration, diffusion, and photophysical effects. FCCS, like fluorescence resonance energy transfer (FRET), can yield molecular interaction information, such as binding. FRAP is used routinely to discover the diffusion and transport properties of fluorescent probes or proteins in the cell, and FLIM measurements can yield a great deal of information on the state of a probe molecule in the cell.

The best resolution one can obtain with optical measurements is on the order of one half of the excitation (or illumination) wavelength. For the 488 nm line of an argon-ion laser, the resolution is therefore roughly 244 nm in the best case. This resolution can be improved upon in the highest-end systems, but not by much. Most of the time, the resolution is worse due to optical aberrations, dirty optics, and so on. Therefore, imaging of the smallest bacteria, virus particles, or ultra-high resolution of cell structure cannot be conducted by routine optical measurements. Fortunately, both AFM and ESEM generate nonoptical images with resolution on the nanometer scale. However, before heading for the local core facility, one must consider if this length scale is important in the desired analysis. The end-user must also recognize that, for the resolution and information content of AFM and ESEM, little chemical information can be obtained relative to fluorescence measurements.

4.3   CULTURING CELLS FOR MICROSCOPY

The methods for preparing cells for microscopic analysis vary greatly by the cell type and microscope. For example, in simple transmission experiments, virtually any transparent material will allow for cell imaging. At high numerical apertures (>0.7), where oil- or water-immersion objectives are used, the thickness of the coverslip often reduces to the order of 150–170 μm. The materials used for coverslip material are low-fluorescence glass. Therefore, a traditional plastic Petri dish or culture flask will not yield proper images. The type of cell also matters. If the cells are alive (and need to remain so during the analysis), then a greater degree of care must be taken. When preparing cells for microscopy analysis, one should follow the following checklist to determine the appropriate method of analysis. Note: In order to avoid confusion, any vessel that contains cells – whether flow is present or not – is referred to in this chapter as a chamber, rather than a "flow cell."

4.3.1   Adherent vs. Suspended Cells

If suspended cells are to be analyzed in a nonflowing environment, they can be placed in the vessel and allowed to settle to the surface or to stabilize in suspension. In the case of a chamber with flow, the adherent cells must be anchored to the chamber surface using fibronectin, poly-L-lysine, or other adhesion-promoting coatings for culture. Adherent cells may either be grown directly on the plate or chamber surface, or can be transferred (see Chapter 3) to the chamber. In the case of adherent cells, it is often easier to grow the cells directly on the surface to be imaged. Several commercial sources sell plastic or glass slides with culture wells attached for easy cell handling. These slides can be of varying thickness for compatibility with dry and immersion objectives, and often are made of

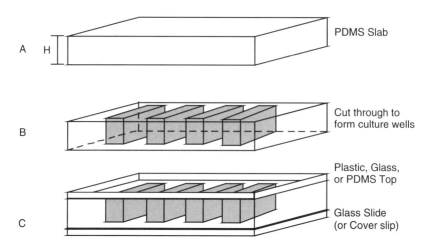

Figure 4.3   A simple, multiple culture well for in-house fabrication. A slab of PDMS silicone (or other suitable silicone) the size of a microscope slide or coverslip (top) is manually cut to house several culture wells (middle, rectangular geometry is shown, although any shape can be used). The silicone piece is then reversibly sealed to the slide or coverslip, and a matching lid made of plastic or PDMS is irreversibly sealed to the top. This device can be reused after sterilization, and the glass surface can be coated with a number of different adhesion-promoting materials (see text)

low-fluorescence materials for minimal background. Plastic slides in this case – which should not be used for sensitive confocal microscopy measurements – are often charged to allow for cell attachment. The glass counterparts are typically covered with poly-L-lysine, fibronectin, or proprietary coatings.

One can also make his or her own culture slides easily in the laboratory. Figure 4.3 shows the general premise. Glass slides – cleaned using successive washes of heptane, isopropanol, and methanol (or ethanol) – serve as the substrate. Other glass-cleaning methods often use strong acids and oxidizers, such as the piranha bath or boiling, concentrated sulfuric acid. These cleaning methods are effective, but should be used with caution. To fashion the side-wall gasket, a silicone material is cut (manually or by a machine shop) into whatever culture-well dimension is needed. The height of the culture chamber is dictated by the thickness of the silicone gasket. Many silicones can be used. Poly(dimethylsiloxane) (PDMS) is an inexpensive choice that is transparent, autoclavable, and biocompatible, but most silicones will also work without issue. The bottom material can be made of any glass or plastic that fits the imaging application. For glass, the surface should be coated to promote cell adhesion. This can be done as described above. One can also, during the coating of the glass with PDMS, coat the entire glass surface with PDMS and cure according to the manufacture's directions (typically 60–70 °C for 30–60 minutes). The culture wells can then be cut into the PDMS slab and peeled away, exposing the glass surface while leaving the walls intact. During this peeling process, nano-scale pieces of PDMS are left on the glass slide [6,7]. This nanoscopic PDMS on the glass surface does not affect the optical properties of the glass, but does allow cells to attach in a manner similar to that in plastic culture ware. Once the materials are chosen and cured, the user simply assembles and sterilizes. The top of the culture dish can be any material. A thin sheet of PDMS allows for gas permeability, but glass and plastic also work well, since gas exchange can occur through the PDMS side walls. Cells must not be left in these slide chambers for several days (even in an incubator), as medium evaporation can wreak havoc on the cells inside. The key advantages to making one's own culture slides are cost, flexibility (i.e., tailoring the culture chamber to the user's needs), and the satisfaction of fabricating something by one's own hands.

If the larger volume of the Petri culture dish is needed, but with the thinness of glass coverslips, one can also purchase dishes modified with coverslip bottoms. A portion (typically 1 cm) of the dish is cut out and a coverslip adhered to the surface. The glass is then coated to promote cell adhesion. These dishes are relatively expensive, but come pre-sterilized for convenience. Again, one can fabricate these dishes using a hole saw and adhering the coverslip to the bottom (Figure 4.4). The drawback of this approach is that the adhesive must be biocompatible *and* compatible with the sterilization methods. For example, many types of glue will soften in an autoclave.

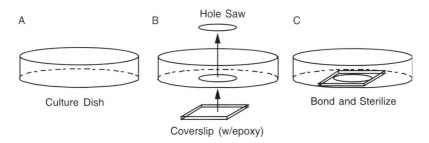

Figure 4.4 Petri dish modified for confocal imaging (adapted from reference [8]). A hole in the bottom of a standard culture dish is cut with a hole saw. An adhesive (capable of withstanding the sterilization process) is used to seal a coverslip to the dish, completing the device. As in Figure 4.3, the glass can be coated with an adhesion promoting compound such as fibronectin

4.3.2   Static vs. Fed Cultures for Viable Cell Experiments

If the cell experiment is of short duration, simply placing the cells in a chamber with a sufficient volume medium will suffice. The environment of the cell may require using a stage heater, but the chamber itself may not require circulation. Cultured lymphocytes in our lab typically can withstand static culture for several hours if the medium volume is large (>500 µl) and the temperature is regulated. However, it is inadvisable to conduct any long-term cell experiment without medium replenishment or chamber temperature regulation. Chapter 9 lists several tests for viability or cell damage that can be used to gauge how long cells can stay in static (no-flow) conditions, or how long they can survive and function without temperature regulation. In the case of the latter, the larger the culture medium or buffer volume, the greater the thermal mass.

Examples of cell tests that could be conducted on any fluorescence microscope include membrane integrity by propidium iodide, 7-amino-actinomycin D (7-AAD), or any other impermeant DNA-binding dye. Viability probes, such as calcein-AM, can also be used to assess both membrane integrity and cell function. A simple approach to conducting these tests is to load a cell sample with the probe(s) of choice and acquire a fluorescence image every few minutes. For light-transmission microscopes, a colorimetric test such as Trypan blue staining would suffice. The same cells are imaged repeatedly and the viability/cell function assessed. It is up to the end-user to decide when viability is too low. In general, static cultures can be used as long as the viability does not drop below 90% of the original ($t = 0$) value.

If the experiment in question requires longer incubation/analysis times, it is best to use a type of flow chamber. Flow chambers vary in size and functionality, but all deliver medium or buffer to the cells. Typically, flow chambers are connected to an external pump and can either be closed loop or open (i.e., drawing from a reservoir and emptying to waste). Medium/buffer flow is provided by peristaltic or syringe pump. The chamber can be open-top (i.e., easily accessed on top by manipulation devices) or closed. Flow rates can vary from static (no flow) to several milliliters of medium per minute, depending on the culture-chamber volume and desired fluid shear force in the chamber. Chapter 7 deals with the effects of shear force on cells; however, it is important to note here that the flow rate must be chosen to allow optimal medium/buffer supply without deleterious effects on the cells. There are many commercial sources for flow chambers, and many designs that can be made in the lab. There is no "correct" flow chamber; rather, one should be chosen that meets the needs of a particular experiment. There are several considerations when choosing flow chambers. First, the issue of open or closed perfusion must be considered. If microinjection, electrodes, or manipulators require access to the cells, then an open system is critical. However, closed-top systems generally provide better sterility control and also minimize blood-borne pathogen risk. Regardless of the type of system, the flow chamber should disassemble easily to clean, sterilize, and replace the coverslip bottom (if applicable).

Heating is also critical for long-term culture. Most commercial systems provide Peltier heaters on the culture chamber, or the microscope stage itself is heated. This approach is challenging, as the objective – especially if oil- or water-immersion – acts as a heat sink. Other culture chambers use indium tin oxide (ITO), , to act as a resistive heating element. There are also noncontact heating approaches, such as hot-air systems that bathe the flow chamber and microscope stage in warm air to control temperature. A less-elegant, but facile approach is to place a small heating element in line between the medium reservoir and the culture chamber. The heating unit elevates the temperature slightly above the desired value, so that when the medium or buffer enters the chamber it is at the correct temperature. This approach has a slower response, but can be implemented by simply wrapping the medium tubing around a small heater or light bulb. A rheostat or variac then controls the power input into the heating element, and a thermocouple registers the temperature at or near the culture chamber. Figure 4.5 shows a simple fluid setup that can be used for either open or closed systems, providing simple temperature control [9]. Control can be manual, or can be linked to computer control software to vary the heating-element temperature.

Many microscopy manufacturers also offer a CO_2 enclosure for culturing using buffers containing carbonate. These enclosures can be small (i.e., fit on the stage) or they can encompass the entire microscope. These enclosures can sometimes be bulky or interfere with other microscope functions. They are, however, sometimes a necessity. The simplest approach to avoiding CO_2 incubation is to use a medium that does not use carbonate as part of the buffer system. If the medium cannot be replaced, then a bubbling chamber can be added to introduce CO_2 gas into the medium (Figure 4.5). A CO_2 sensor would ideally be placed in line to monitor gas readings; however, if the system is stable, then a one-time reading could be used to calibrate the CO_2 content in the medium or buffer. Another issue with flow chambers is the introduction of bubbles. A

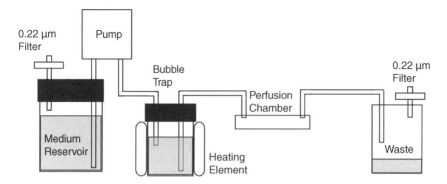

Figure 4.5 Fluid system for perfusion of medium for live cell microscopy. The perfusion chamber (see text) is mounted on the microscope stage. Medium is drawn from a reservoir by a peristaltic or syringe pump. A bubble trap in the fluid line eliminates bubbles while a controllable heating element adjusts the medium temperature to physiological temperature (if the bubble trap is omitted, the reservoir can be heated). The 0.22 mm filters on the reservoir and waste eliminate pressure buildup. This approach is simple, modular, and does not require a stage heater in many cases. A variety of perfusion chambers can be used

bubble, however small, can cause flow problems or, at worst, dislodge cells from the surface [10]. A bubble trap (Figure 4.5) is a simple device that would be placed after any CO_2 sparge chamber.

Culturing cells for AFM varies little from optical measurements (indeed, many AFM systems can be integrated with optical microscopes). Since the AFM cantilever must access the cell sample, an open-top system must be used. For ESEM measurements, the sample is placed in the imaging chamber and the pressure is kept higher than typical SEM applications, but much lower than atmospheric pressure. The 10–20 torr pressure is sufficient to maintain a moist atmosphere around the cell. The end result is that cells do not desiccate as much as they tend to under high vacuum. Also, the presence of some residual water eliminates the need to sputter coat the sample (for conductivity reasons).

By necessity and design, it is not currently possible to maintain homeostatic conditions during ESEM imaging. The ESEM process is considered an invasive technique in which cells are killed during the measurement. However, the spatial resolution – and the ability to perform electron immunomicroscopy – outweigh this constraint. For ESEM culturing, cells should be either grown on or attached to a suitable substrate. It is best to contact the SEM facility (if a core instrument is used) to ascertain what samples (and sample sizes) can be accommodated.

4.4   SIGNALS, BACKGROUND, AND ARTIFACTS IN OPTICAL MICROSCOPY

To the inexperienced user, cell analysis by microscopy often appears to be an art more than a science. Indeed, often younger graduate students are frustrated when an older lab mate tells them the cells "just look right," or "clearly are dead" without the use of a stain. Experience does, however, play a large role in how easily one can observe a cell sample by eye and tell the relative state of the sample. A desiccated cell sample looks much different than a fresh one. A live cell can be distinguished from a long-dead cell by morphology alone. Morphology, unfortunately, is highly subjective, and subtle differences are easily lost to most observers. The best cell analyses remove as much subjectivity as possible. There is already a built-in bias when the user moves the microscope stage around looking for cells. Despite the best of intentions, cells that do not "look right" are sometimes unintentionally – and unfortunately sometimes deliberately – passed over for more favorable-looking cells. To eliminate this bias, one should either randomly set the microscope stage, or measure as many cells as possible to reduce user bias.

In transmission or DIC microscopy, contrast is generally provided by differences in the refractive index of the cell and the surrounding medium. The difference in refractive indexes is generally small. A dead cell, for example, will have a compromised membrane. Freshly dead cells retain much of the cytosol and appear, to the eye, viable. A long-dead cell, however, shows little contrast compared to its healthy neighbors. This lack of contrast is one the tricks that the experienced observer knows and exploits. Of course, for accurate viability measurements, one should use a membrane-impermeant dye (Chapter 9), rather than rely on a visual estimate.

Apoptosis is another example of morphology and contrast used to assay a complex cell mechanism. The blebbing of cells in the late stages of apoptosis is used as proof that a cell has undergone programmed cell death. However, there are many fluorescence and immunogold micro-scopy techniques (Chapter 9) for imaging cells and identifying them in various stages of apoptosis. Morphology and contrast are useful, but they should be used in moderation, as they tend to be subjective and not sensitive. Sources of error in contrast measurements include ghost cells (the shells of long-dead cells), out-of focus cells and particles, and the detritus that can at times accumulate as cells proliferate and die.

In fluorescence microscopy, one of the main sources of error and artifact is background fluorescence. This type of signal can occur from three main sources. First, the cell itself fluoresces to some extent at certain wavelengths. Second, the culture medium or buffer may fluoresce, making cell imaging difficult. Third, debris in the sample can also fluoresce – in some cases much brighter than anything stained in the sample. Autofluorescence typically occurs in cells excited at wavelengths lower than 550 nm (this cutoff varies from cell type to cell type). The simplest test for autofluorescence at a given excitation–emission wavelength is to image the cells under the appropriate measurement conditions (exposure time, etc.) in the absence of any fluorophores. Cell autofluorescence is worse at shorter wavelengths, which makes blue-emitting dyes difficult to image. There are several mechanisms to mitigate autofluorescence, and each has its own merit, depending on the experiment at hand.

If a fluorophore is added to the cells after the sample is mounted onto the microscope, it is possible to photobleach the autofluorescence of the cells beforehand. It is important to stress that this preliminary photobleaching step should be performed in the absence of any dye, antibody, or other fluorescent probe. Photobleaching in this case requires exposure of the cells by the excitation source. This exposure can occur at higher intensity (to shorten the bleaching time), but care must be taken to avoid phototoxicity and heating effects. This photobleaching time and power should be deduced carefully for the cell types under investigation. The simplest mechanism for determining a photobleaching protocol is to expose a control sample of cells at a desired power and time, then incubate the sample with a membrane-impermeant DNA-binding dye such as propidium iodide, 7-AAD, or Sytox green. Comparison to nonexposed cells will allow one to therefore determine a cutoff power and exposure time. Fluorescence images should then be acquired at regular time intervals for what would be the entire duration of the desired cell experiment, to ascertain the long-term effect of photobleaching on cell viability. For reliable measurements, one should operate well below the established cell-damage threshold. Note that this analysis should be performed for each microscope magnification, wavelength, and cell line used. Once the exposure is determined, however, those values may safely be used until the experiment or microscope change. For fixed cells, viability and function are unimportant, but care must also be taken to avoid heating damage and morphological changes.

It is also important, during the optimization of the photobleaching step, to assure that autofluorescence recovery is minimal. Monitoring cell fluorescence after bleaching at several time intervals (that match the experiment timeframe) readily establishes the recovery rate. Another factor that must be considered is that viability alone is not the only side-effect of a large light dose. Cell morphology and function should both be tested. Calcein-AM is a versatile dye for determining morphology, membrane permeability, and general cell activity. Again, the dye should be added after the photobleaching step. For specialized measurements, such as fluorescence correlation spectroscopy, photobleaching may be necessary to remove uncorrelated autofluorescence before the correlated fluctuations of freely diffusing species is measured [11]. Initial cell measurements do not show autocorrelated fluorescence, as rigidly held autofluorescent materials obscure any diffusion-related fluctuations. This effect is problematic in confocal microscopy of single (or few) molecules, as the out-of-focus light zone is significantly larger than the confocal region. Despite the reduced detection efficiency of out-of-focus light, the large size contributes to poor autocorrelation functions. Once the auto-fluorescent material (and analyte) is bleached away, free dye can now enter the probe volume and generate a correlated fluorescence signal.

In multi-photon excitation microscopy (MPEM), two or more photons of longer wavelength are simultaneously absorbed for a transition that is of a shorter wavelength. One of the many advantages to MPEM is that excitation only occurs in the focal plane. Autofluorescence, therefore, is restricted to the focal plane only. While MPEM techniques general enjoy a lower background signal, the current cost of ownership of such a system is higher than other microscopy systems. Therefore, MPEM should be viewed as a technique that – fortunately – has lower background, rather than a technique to *pursue* to reduce background. Another approach to reducing autofluorescence is to either use a brighter fluorophore or move to a longer wavelength. This is not always possible, as is the case of most calcium-signaling dyes or rhodamine-based fluorogenic probes. However, if the fluorophore can be substituted, moving to a longer wavelength, such as a dye excited at 635 nm, typically results in a higher signal-to-background ratio. An added benefit of moving to red-excited dyes is that diode-laser excitation is significantly less expensive than other laser sources.

A subtler source of background is from the culture medium itself. For most epifluorescence and confocal measurements, the background fluorescence of phenol red (added to most formulations of cell-culture medium) obscures the desired analyte. Phenol-red fluorescence is problematic for many green-emitting dyes and fluorescent proteins. For sensitive measurements by single molecule detection, FCS, or FRET, the protein components of the medium also pose a challenge. Serum, typically added at 5–10% concentration, causes a significant background for green fluorophores excited by the 488 nm line of an argon-ion laser. For short-term measurements, switching the medium with a buffer results in background-free imaging. For longer-term culturing on the stage, it is best to culture the cells with phenol-red-free medium. It is best to determine the effects of the culture medium in the absence of cells, and then optimize the cell experiment itself.

Debris in or out of the image plane can create a large background signal. Fortunately, most debris (dust, probe aggregates, etc.) are readily distinguished from the cell sample by shape and size alone. However, if a dust particle looming above the cells poses a problem, it is best to find a new field of view that is free of debris.

4.5 STAINING CELLS FOR FLUORESCENCE MICROSCOPY

Similar techniques can be used for staining cells by transmission or fluorescence microscopy. However, since there are stricter protocols for fluorescence work, this more conservative approach will also apply to transmission measurements. Unlike flow cytometry or other measurements where the cell has a transient residence time in the excitation beam, staining cells for fluorescence microscopy is more complicated, since the cell itself is stationary. This means that the fluorophores bleached during observation cannot be replenished. In flow cytometry, for example, transit times in the laser beam are 1–$1000\,\mu s$, depending on the system used. The cell then passes to waste, or – in the case of sorting instruments – can be collected for further study. In microscopy, however, the cell is isolated in space and irradiated for a significant amount of time. Shorter exposure times last several milliseconds, and cells are routinely observed for minutes or even hours, although light exposure does not necessarily occur for that entire period of time. The irradiance of the excitation light is high in order to excite even weak fluorophores. As a result, photobleaching is often a rapid and unavoidable process. Even in the case when a cell is imaged briefly, but repeatedly, for a long period of time, photobleaching of the probe is significant. Often, the act of finding a cell and focusing by fluorescence diminishes the fluorescence signal significantly. In addition to finding and focusing on the specimen, one must also optimize the charge coupled device (CCD) camera exposure time, gain (if any), and so on. By the time an actual analytical experiment is made, the cell may be too dim to image. One can, of course, optimize measurement conditions on one field of view, and then move to a fresh area on the sample to image cells. Using the microscope, lamp, or laser shutter is critical in this case, so that cells are only exposed to excitation light during the actual measurement.

In the case of a sample containing few cells of interest, it is impractical to destroy one of the few cells to be measured in order to set up instrument parameters. In these cases, it is better to use microspheres – also referred to as beads – to set up the microscope. Fluorescent microspheres are sold in a variety of sizes and fluorophores, with brightness that can match most cell samples. A vial of beads matching your cell analysis is a worthy – and small – investment compared to the rare cell. Another approach is to keep a sample of fixed cells of similar fluorescent properties in the refrigerator or freezer for calibration. Again, rapid control of the excitation shutter is critical to preserve signal.

One of the major reasons that photobleaching is problematic in microscopy is that cells are typically observed over long periods of time. There are two main causes for the long observation times normally used. First, the cellular process itself may be a slow one that lasts for hours. Second, for even a "fast" measurement, a longer exposure may be needed to obtain an adequate signal-to-noise ratio. If the imaging setup is shot noise limited, and the CCD detector used reads out in photon counts, then the signal-to-noise ratio will increase with the square root of the image acquisition time. Even in this best-case scenario, the square-root dependence of signal-to-noise on the acquisition time means that there is a point of diminishing returns on image quality. For example, Figure 4.6 shows an image of a mouse endothelial cell expressing green fluorescent protein. Ten short (20 ms) exposures were made of the same cell. The signal-to-noise ratio in the first image is about 3.8. Each additional exposure that is averaged produces a higher signal-to-noise ratio. However, after image #6, the increase in signal-to-noise is small and the analysis time (and therefore degree of photobleaching) now detracts from the incremental improvement in performance. It is best, therefore, to optimize each measurement period for the best, practical signal-to-noise ratio. If a better, that is, more sensitive, detector is used, the exposure time can be shortened while maintaining a good signal-to-noise ratio. The same applies to higher-efficiency objectives and optical filters. The exposure for a "fast" measurement can be well less than a second for a bright fluorophore at high concentration, whereas a weakly stained cell may require several seconds, even with a sensitive CCD detector.

For long-term cell processes, it is best to critically examine the temporal resolution needed to image the sample. For example, a cellular process that takes two hours can probably be imaged with high temporal resolution if a short (e.g., 1 s) exposure image is taken every 30 s. The resulting data file would contain 240 images and would have exposed the cell to excitation light for four minutes. It is unlikely that the fluorescence would

(a)

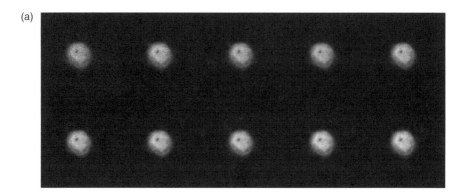

(b)

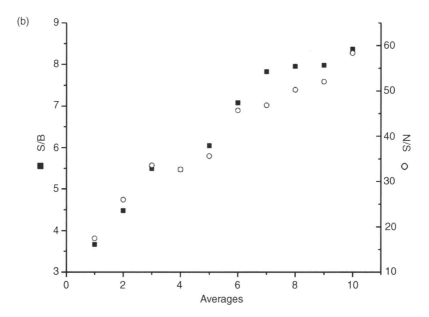

Figure 4.6   Fluorescence microscopy images (a) of a embryonic mouse endothelial cell expressing green fluorescent protein (GFP). Each image an average of n successive 20 ms exposures ($n = 1$–10). With each average (b) the signal-to-noise (open circles) and signal-to-background ratios (black squares) increase. However, after about six averages, in this case, the improvement in image quality is negligible, while photobleaching increases

recover in a cell with this type of exposure. One could, therefore, shorten the exposure to the minimum necessary to observe an adequate signal-to-noise ratio, and then take measurements with a low enough duty cycle to avoid significant photobleaching. In a two-hour process, half-minute images might already oversample the temporal process at hand. The best way to determine the temporal resolution needed is to understand the mechanism that is being studied. In caspase activation (during apoptosis), fluorogenic and caspase probes are fully cleaved within 15 minutes of caspase activity [12,13]. The process of caspase activation – from initiation to cleaved substrates – can occur at longer timescales. To determine caspase activity over the course of an hour, one image every minute would suffice to not only observe the delay between apoptosis induction and caspase activation, but also the kinetics of the probe cleavage. Switching to an image every two minutes would still provide adequate resolution, although less so. Taking images more than once a minute would not likely yield any additional information, but would increase the degree of photobleaching.

One of the most overlooked solutions to minimizing photobleaching is to optimize the excitation light. Most microscope users operate the lamp at the normal – maximum – output. However, the irradiance from most lamps is significantly higher than what is needed for microscopy. A simple numerical example illustrates this. Fluorescein, one of the most popular fluorophores, has a lifetime of 4.5 ns in the absence of quenchers. This means that the time between excitation – which occurs essentially instantaneously – and emission is, on average, 4.5 ns. One can calculate that if a photon arrives every 4.5 ns, a fluorescein molecule will have been excited from and returned to the ground state. In this case, the molecule is once again ready to absorb a photon as it arrives (the probability that the molecule will absorb that photon while in the excited state is zero). If the average photon arrival time at the sample is more than 4.5 ns, then the molecule essentially waits until the photon arrives, which is not the most efficient approach [14]. Therefore, an increase in the excitation irradiance (power per unit area in the focal plane) would result in an increase in fluorescence intensity up until the point that the photon arrival rate equals the fluorescence lifetime. Once this condition is exceeded, photons arrive more rapidly, and the period between photons is less than 4.5 ns. During that time, the fluorescein molecule is in the excited state and working toward relaxing to the ground state. These additional photons are not absorbed. Of course, neighboring fluorescein molecules will absorb the additional photons, but as the irradiance increases they too will occupy the excited state while additional photons arrive. At this point of optical saturation, there is a decrease in the observed absorption and – more importantly – no increase in fluorescence. There are some distinct advantages to saturation conditions (less noise, quenching effects are reduced, etc.). For cell imaging, however, photobleaching increases beyond saturation with no net increase in signal. Therefore, for optimal cell imaging, the power should be increased until fluorescence just stops increasing. Working beyond this point only degrades the sample and destroys the fluorophore.

To decrease the excitation irradiance, the simplest approach is to invest in a set of neutral density filters for the microscope. The power can then be attenuated by a range of values in order to reduce photobleaching while maintaining a high signal. For many commercial laser confocal microscopes, the laser power can be attenuated by acousto-optical filters or polarization optics for a nearly continuous power range. In this case, imaging conditions can be optimized more readily.

Photobleaching of fluorophores is especially problematic in confocal microscopy. Most microscopic methods are essentially 2D, and the entire cell can be imaged simultaneously in one exposure. In laser- (or stage-) scanning confocal microscopy, the beam raster requires a longer exposure when 3D images are acquired. Figure 4.7 shows the confocal volume in a cell, as well as the out-of-focus volume above and below the focal plane. While the excitation irradiance is highest in the focal plane, photobleaching in the out of focus region is significant. As the 3D scan progresses, the entire cell is exposed repeatedly to the scanning laser. A simple test to determine the extent of photobleaching in confocal measurements is to

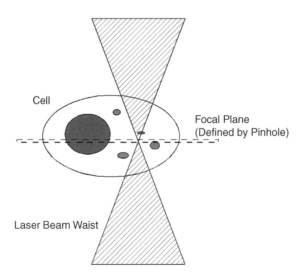

Figure 4.7 Diagram of the laser beam inside a cell in confocal microscopy. The confocal pinhole optics restrict detection to the focal plane. While excitation is highest in the focal plane due to the smallest laser spot size, excitation – and photobleaching – by the laser in the out-of-focus region is ever-present

rapidly measure the bottom of the specimen, then scan from the top of the cell downward, comparing the final intensity of the bottom frame to the initial scan. As in conventional light microscopy, better detectors, stable fluorophores, and the optimum laser excitation power will minimize the effect of photobleaching.

Photobleaching occurs through several processes, including reactions between the excited state fluorophore and oxygen species. For fixed cells, it is possible to add oxygen-scavenging species that reduce photobleaching. There are several reagents available commercially, including the Prolong and Slowfade products from Molecular Probes (now part of Invitrogen). Most anti-fade reagents are not amenable to live-cell imaging, so care must be taken to ensure biocompatibility if live-cell imaging is required. In addition, many of these mountants require some level of hard-setting. The antioxidant ±-6-hydroxy-2,5,7,8-tetramethylchromane-2-carboxylic acid (trade name Trolox) is a common anti-fade reagent that can be used with live cells and does not require hard-setting or nonaqueous conditions.

Table 4.1 lists many types of artifacts/obstacles in cell imaging by microscopy, as well as several steps to evaluate and avoid them. Not every method to reduce interferences and artifacts can be employed in every setting. However, there are several steps that can be taken for most experiments to reduce background and maximize signal.

Minimizing photobleaching of a dye, as well as autofluorescence and background signals are just some aspects of image optimization. The signal itself also plays a critical role in the successful cell analysis. Most fluorescent signals can be split into two distinct categories [15]: probes and labels. A probe is a fluorescent molecule that changes some spectral property (intensity, polarization, wavelength, lifetime) upon interaction with the environment. An example would be a pH-dependent probe such as SNARF that changes intensity as pH changes. Calcein-AM and other fluorogenic reagents also are classified as probes, since enzymatic activity changes the fluorophore properties (from nonfluorescent to highly fluorescent). Labels, on the other hand, are defined by their presence or absence. An example of this would be GFP plasmid constructs that report gene expression. GFP itself does not interact with a cellular component or chemical species, but "appears" as a result of gene expression. Antibodies are another excellent example of a label. In the case probes, one is often limited in choice of wavelength and photostability. Most calcium indicators feature blue emission and UV excitation. Fluorogenic rhodamine probes span the 500–600 nm range, but are not widely available for red excitation. With labels, there is a greater degree of flexibility. For a given antibody, one can often find at least 2–6 different fluorescent labels, as

Table 4.1 Troubleshooting microscopy

Problem	Troubleshooting
Image is blurred (for immersion objectives)	Glass/coverslip too thick Immersion medium absent (or incorrect immersion medium)
Contrast is low (transmission microscopy)	Cells are dead or fixed. Use DIC or sample with higher viability
Cells are dead (and bacteria present)	Sample is contaminated. Use closed perfusion/observation system; add antibiotics to medium
Cells are dead (no contamination present)	Use a stage heater (set to 37 °C) Perfuse medium
High fluorescence background	Use phenol red-free medium for imaging Use culture ware with low fluorescence Photobleach background (if possible)
High autofluorescence from cells	Photobleach autofluorescence before probe/label added Use longer-wavelength probe/label
Fluorophore bleaches during analysis	Use shortest possible exposure time (better detector) Use lowest possible excitation power Use the minimum number of averages Optimize temporal resolution Minimize Z-scanning (confocal) or use TPEM
Image is noisy	Using image averaging (but minimize photobleaching and optimize temporal resolution)
Fluorophores bleed through detection channels	Use longer Stokes shift dyes Use alternating excitation (if possible) Use shorter wavelength dyes for the weaker signals
Poor Z-resolution (confocal, thick sample)	Use TPEM/MPEM Use the correct coverslip thickness
On-screen image is too dim or too bright	Use a brighter monitor Maintain brightest signal below saturation (2^n bits)
Poor lateral (XY) resolution	Increase numerical nperture (NA) Use smaller-pixel CCD (affects dynamic range) Use confocal or TPEM/MPEM
Cell ruptures during analysis (AFM/ESEM)	Use higher imaging chamber pressure (ESEM)
Poor cell morphology (AFM/ESEM)	Use noncontact (tapping) mode (AFM)

well as biotin. Quantum dots, heralded for several years as a replacement for organic fluorophores, can be used as luminescent tags for labels, such as antibodies.

If a probe is used, one must ensure that the change in the dye is readily observed in the system of interest. A fluorogenic probe, for example, will have limited sensitivity if the fluorescence of the cleaved probe overlaps with strong autofluorescence. DNA probes offer a wide spectrum of options from blue to far-red excitation. Probes for pH and metal ions (Ca^{2+}, Mg^{2+}, etc.) also offer a large degree of flexibility (see Chapter 9 for specific examples).

Maximizing signal and avoiding photobleaching are just two of the challenges faced when staining cells. Investigating cells using labels or probes requires access to the cell surface, its interior, or both. In the case of cell-surface measurements, staining is relatively straightforward (Chapter 9). For longer-term measurements, or for surface staining where the label can detach, it is best to fix the cell. Most fixatives for staining are protein cross-linkers such as aldehyde-based reagents. Cell fixation can be a complex and oftentimes error-prone process. The essential steps to fixing the cell are as follows: washing, fixing, and washing again. The first washing steps eliminate protein-containing materials in the buffer or medium that could be cross-linked to the cell. The label can be attached before this step, as any additional label would be washed away. Formaldehyde and paraformaldehyde are the two most common fixatives, and are usually prepared as working solutions just before fixation. Neutral buffered formalin is also available commercially for cell fixation. All aldehyde-based fixatives age over time, so it is best to store the fixative as $-20\,^{\circ}C$, or to prepare freshly, to maintain potency.

PROTOCOL 4.1 FIXATION OF CELLS FOR IMMUNOCHEMICAL STAINING

While fixation protocols vary by application, and should generally be optimized for new experiments, a straightforward protocol for formaldehyde fixation is listed below (adapted from [16]).

1. If the cells are suspended, centrifuge three times, resuspending in phosphate-buffered saline (without Ca^{2+} or Mg^{2+}, pH 7.4; PBS). For adherent cells, decant the medium and wash with PBS three times for five minutes each.

2. If cell staining occurs after fixation, skip to step 5.

3. If the cell surface is to be stained before fixation, incubate the label in buffer with 5–10% bovine serum albumin (or other incubation buffer) for the desired time.

4. Wash three times with PBS.

5. Replace the medium with 10% formaldehyde in PBS for 20 minutes at 4 °C.

6. Wash three times with PBS. Mount with PBS or glycerol-based medium and analyze.

7. If the cells are to be stained after fixation, add the labels at this point and incubate according to step 3 (or manufacturer's protocol).

8. If permeabilization is required, steps 1–7 of the proceeding protocol are followed by permeabilization of the cells with 0.1% Triton X-100.

For most probe measurements, cell fixation is not desirable, as the act of fixation alters the environment that the probe is measuring. It is best, therefore, to choose a cell-permeable probe. Most probes, therefore, are smaller molecules with a fair degree of cell permeability. Of course, if the probe is measuring membrane integrity, then high permeability is not desirable (except for some viability probes that have a high initial viability and are converted to an impermeant form). The issue of permeability becomes an obstacle when intracellular antigens are measured. Antibodies, antibody fragments, aptamers, and other labels are too large to freely enter the typical mammalian cell. Intracellular labeling agents can be introduced into the cell by several formats. If viability is not required, then cell fixation followed by permeabilization will efficiently allow transport of large labels into the cell. Fixation by formalin, followed by saponin or Triton X-100 permeabilization, allows intracellular antigen staining to occur with minimal investment.

For cell measurements where the cell must remain viable, but a large probe must be delivered, several options exist. Biological delivery agents are available to deliver proteins and other reagents to cells. They are typically either liposome or peptide based. A short, repeated arginine (poly-arginine) sequence can enhance cell permeability. Mechanical methods, such as electroporation and microinjection, effectively deliver reagents to live cells. Electroporation is sometimes a more efficient technique, and it has been shown to transfect difficult cell types [17,18]. The electric fields used in electroporation can lyse cells and also create a backflow gradient [19]. Microinjection has the additional benefit that only particular cells are loaded with the reagents of interest. Of course, if many cells need to be loaded, a more uniform approach, such as electro-poration would be a better choice.

For microinjection, cells must be anchored on the substrate to prevent movement. Another approach often used for oocyte microinjection is to gently hold the cell in place with a micropipette pulling a slight vacuum. For most other cell analyses, however, anchoring the cell is sufficient (and simpler). Cells are incubated on washed glass slides treated to promote adhesion. The cell type and conditions must be optimized for each experiment – once again, a degree of optimization must occur before routine analyses can be employed. If one considers a typical lymphocyte to be a sphere with a diameter of 15 μm, the volume of the cell is approximately 2 pl. Delivery of reagents therefore must be precisely controlled to avoid rupturing the cell. During microinjection, the most common causes of failure are as follows:

1. Pressure too high/injected volume too large (the cell ruptures).
2. Tip crashes into coverslip and breaks.
3. The cell distorts significantly because of large injection. Some minor cell distortion is not indicative of a failed injection, particularly if the cell recovers after a short time (1–10 minutes). However, significant distortion indicates a compromise cell.

4.6 MULTIPLE LABELS

So far, the case when only one label or probe is added has been addressed. In immunochemical imaging, often more than one probe is measured, complicating the analysis. The rules for photobleaching of the background (and the fluorophores) still apply. What one must now consider is the effect of signal bleed-through and interference between probes. Signal bleed-through is problematic for many single wavelength-excitation systems. Indeed, a microscope exciting at 450–490 nm with a 500–540 nm emission filter is limited to green-fluorescing dyes such as fluorescein, rhodamine 110, and green fluorescent protein. The overlap of these dyes is such that only one can be used at a time. However, fluorophores with longer Stokes shifts (differences in emission maxima), such as R-phycoerythrin can be used with one of these green-fluorescing fluorescent molecules. Longer-shifted dyes based on metal ligands such as europium and lanthanum also extend the range of fluorophores and minimize overlap. Figure 4.8 shows an example of the overlap between several dyes excited with a 450–490 bandpass

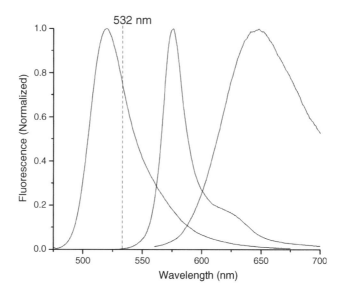

Figure 4.8 Emission spectra of fluorescein, R-phycoerythrin, and 7-aminoactino-mycin D (7-AAD), as well as the emission line of a 532 nm Nd:YAG laser (dotted line). The overlap between dyes requires compensation, particularly if wide-band interference filters are used. The laser line is shown to emphasize that in multi-laser systems, the excitation of one laser may interfere with the emission of a dye excited by another laser

filter and a 100 W Hg lamp. The overlap of signals depends on the quality and bandpass of the emission filters and dichroic mirrors used in the microscope. For flow cytometry, fluorescence overlap is minimized by a process called compensation (see Chapter 6 for a discussion on compensation).

As shown in Figure 4.8, the best way to avoid overlap is to use two fluorophores with a large Stokes shift. However, in some cases there is a limit to how well-separated these peaks are. When more than one excitation wavelength is used, the Stokes shift is typically considerably larger. For confocal instruments, both lasers can be used simultaneously or sequentially, and detection of both emission wavelengths (or more) occurs at the same time. Therefore, the analysis time is not increased (if both excitation and emission sources are on at the same time). In this case, however, one must ensure that each excitation line does not overlap with the emission band of the proceeding dye. For example, in a confocal system that contains a 490 nm/532 nm solid-state laser, the scatter from the green (532 nm) laser overlaps with the emission of most green fluorophores (see Figure 4.8). In this case, the lasers must be alternated, as the scattered signal will overshadow the fluorescence in almost every experiment. For light microscopes using broadband lamps and excitation filters, each excitation must occur sequentially regardless of overlap. For multiple excitations and exposures, the total light dose to the cell increases, and the overall analysis time increases by the number of excitation/emission wavelengths imaged. There is therefore a practical limit to the number of fluorophores imaged by conventional epifluorescence microscopy that is not experienced by confocal instruments (which are typically limited by the number of detectors).

In order to gauge the effect of signal crossover between channels, $N + 1$ control samples should be evaluated for a given set of experimental conditions (where N is the number of fluorophores imaged). One sample will be completely unstained to evaluate the degree of autofluorescence from the cells and medium/buffer at each fluorescence channel. The remaining controls will each be stained with just one of the probes/labels. All $N + 1$ samples are then imaged at each wavelength, and the relative signal recorded. Ideally, the signal should be high in the desired channel, and zero for all remaining channels. However, one quickly observes that with smaller Stokes shifts, or intense staining, the bleed-through between channels is large. If the cross-talk is small, then imaging can continue without significant error. However, bleed-through of more than a small percentage can be mathematically subtracted (See Chapter 6).

In *Practical Flow Cytometry* [15], Shapiro points out that the overlap of the shorter wavelength fluorophores into longer wavelength channels in flow cytometry is typically greater than bleed-through in the opposite direction. Therefore, it is best to use shorter-wavelength fluorophores for labels/probes of lower signal. For example, if one is trying to ascertain the viability of a cell with calcein-AM, and also label a cell surface antigen with an antibody conjugated to R-phycoerythrin, the overlap of the intense green fluorescence would create a significant error in the measurement of the antibody. Using a viability probe at a longer wavelength and using a green fluorophore, such as AlexaFluor 488, for the antibody, would result in less of a cross-talk between the fluorescence channels. This approach also applies to confocal measurements, two-photon excitation, and flow cytometry.

4.7 VIABILITY AND TWO-PHOTON MICROSCOPY

Multi-photon excitation and two-photon excitation (MPE and TPE, respectively) are powerful tools for imaging of cells, cell aggregates, and tissues. As mentioned above, MPE and TPE only excite fluorophores of interest when the photon irradiance (photons cm^{-2} s) reaches a threshold value. In confocal, one-photon excitation microscopy, the probe volume is somewhat loosely defined as a Gaussian ellipsoid defined by the beam waist on the confocal pinhole. In MPE/TPE, the pinhole is removed as the excitation volume truly is an ellipsoid with a sub-femtomolar volume. Therefore, in the vast out-of-focus region of the laser beam, little to no excitation occurs. MPE/TPE techniques therefore suffer from less photobleaching than their one-photon counterparts. Indeed, the scan times can be much longer in MPE/TPE microscopy since only the focal plane experiences any significant bleaching. However, one must consider that multi-photon excitation is a nonlinear process. In order to efficiently excite a fluorophore with two or more photons, the instantaneous irradiance must be high. Picosecond and femtosecond, mode-locked lasers are the sources of choice for MPE/TPE applications, particularly titanium–sapphire (Ti:Sapphire) mode-locked lasers. The Ti:Sapphire is broadly tunable (690–1080 nm), allowing different fluorophores to be excited. Whichever laser is used, however, one must consider that while the two-photon photobleaching and phototoxicity is limited to the focal spot, one-photon effects are not.

The high average (and pulsed) power of these mode-locked lasers requires that the effects of photobleaching and phototoxicity be carefully studied for a given sample. In most cases, the long excitation wavelengths (typically >800 nm) do not result in significant photobleaching of the sample. Heating, however, is a much more prevalent problem [20]. The effect of the incident wavelength, at high power, can heat viable cells to the point of death, and can also distort and deform fixed cells.

Unlike conventional one-photon excitation, where optimizing excitation power is relatively straightforward, adjusting the excitation power in MPE/TPE is more difficult. For TPE, there is a square dependence on the laser power; for three-photon excitation, the dependence is to the third power. Therefore, to decrease the *rate* of excitation by 50%, one would have to lower the laser power by 30% for two-photon excitation and by approximately 20% for three-photon excitation. The ability to control the laser exposure is more difficult with fluorophores that excite weakly by MPE/TPE. Just as in one-photon excitation, it is best to irradiate different cell samples with increasing laser power, and then monitor for cell viability. In the simplest approach, a mercury lamp should be combined with the mode-locked laser, so that cells irradiated during the test scan can be rapidly assayed using a viability probe and epifluorescence imaging. Just as in one-photon excitation, one should assay a range of power and exposure time.

4.8   SPATIAL RESOLUTION IN OPTICAL MICROSCOPY

When choosing a microscopy technique, both the chemical, biological, and spatial information needed must be considered. As discussed in Section 4.2, fluorescence microscopy offers contrast and chemical selection that transmission microscopy cannot. However, when considering spatial information, all optical microscopes have similar features, with the exception of confocal microscopy. When one considers the spatial resolution needed for a given experiment, the constraint on optical performance increases. Indeed, at times, the required resolution precludes optical measurements altogether. In Figure 4.1, the resolution needed for a given analysis may dictate what type (or what quality) of microscope is chosen. For example, imaging mammalian cells, with little interest in organelle structure beyond the nucleus, will require much lower resolution than bacterial imaging. When considering resolution in cell imaging, a distinction must be made between the distance between two spots (i.e., the feature to be imaged) and the resolution element of the microscope, called the resel. The minimum resolution of an optical microscope under wide-field imaging is given by the following equation:

$$R_S = \frac{0.61\lambda}{NA}, \qquad (4.1)$$

where NA is the numerical aperture and λ is the illuminating wavelength. For confocal microscopes, the reduced field of view improves R_S by the square root of two. Thus, for a typical microscope using a $20\times$, 0.40 NA objective, the minimum resolution would be approximately 750 nm with 500 nm light. Switching to a $60\times$, 0.67 NA objective under the same conditions would improve the resolution to 450 nm.

The minimum feature that can be resolved will be twice the resolution element (i.e., $2R_S$). The factor of two for R_S stems from the need to measure any sampling event by at least two intervals, according to the Nyquist theorem. In frequency terms, a 1 kHz sound wave would have to be sampled with a sampling rate of at least 2 kHz to faithfully reproduce the original sound. For measurements of distance, the sampling interval should be $d/2$, where d is the distance to be measured. It should be stressed, however, that the Nyquist theorem outlines the minimum resolution needed. In fact, sampling at the Nyquist limit may yield unsatisfactory results, as subtle changes or other factors (such as the shape of an organelle, not just its size) will require a higher sampling "rate." For example, measuring apoptotic blebs by microscopic examination requires a resolution necessary to observe the bleb size, but also to distinguish neighboring blebs and general debris in the image plane. For most mammalian cell analyses, conventional microscopes with $20-60\times$ magnification objectives will provide sufficient resolution. For higher resolution imaging of organelles and bacteria, objectives in the $40-100\times$ range, with high numerical aperture, are required. The final resolution also depends on the microscope detection. Since CCD cameras have largely replaced film cameras on microscopes, the CCD pixel size also plays a role in the resolution. In the previous example of a spacing of 450 nm, the image formed by the $60\times$ objective would not be resolved if it were projected onto less than two CCD pixels. Therefore, the magnification must be sufficient to project the resolved distance onto the camera. Using this example, the final projected distance of the 450 nm separation (R_S) is 27 μm. In this example, as long as the CCD pixel size is less than 13.5 μm, the resolution obtained by the objective is preserved. It is important to note that, as mentioned above, the two-pixel minimum is the absolute lowest sampling interval; for most work more than two pixels (or resels) should be used.

In order to make distance-based measurements of cells (e.g., length/width, depth), the microscope must first be calibrated. Inexpensive, chrome-on-glass resolution targets for microscopy are a straightforward approach to spatial calibration in the x–y plane of the microscope. Ronchi line targets, with 600 lines mm^{-1} yield micron-scale resolution targets to test most microscopes. The USAF 1951 resolving target is also useful (and can be used to identify microscope astigmatism), but the group number should be large enough to permit imaging at $60-100\times$ objectives. There are a variety of calibration slides/plates, made of glass or fused silica, or opal glass. The masks can be positive or negative, and can have fluorescent material applied as well. Most resolution targets have features that range from 1 μm to 1 mm; however, when purchasing or using, one must verify the range of the mask features. For depth (z-axis) resolution on confocal instruments, the precision of the depth measurement depends largely on the step accuracy of the z-axis stage. It is difficult, therefore, to calibrate the depth by optical methods. Polystyrene beads of low size dispersion can be used to calibrate the z-axis measurement, but the resulting accuracy will be lower than a chromed mask.

One of the main reasons that the laser beam is typically rastered (rather than the sample) in laser-scanning confocal microscopy is that the sample itself should not move during analysis. In most cases, especially with adherent cells, sample movement (even in the presence of perfusion) is negligible, if present at all. However, for suspended cells if anchoring steps or low-shear perfusion devices (see Chapter 7) are not used, the cells can move on the stage. Even in the absence of flow, cells may move as the liquid volume changes due to evaporation, or by optical effects. Laser trapping/ manipulation of particles has become a useful method for moving cells, but trapping effects can cause displacement even in standard confocal instruments. As laser power increases, the ability to move cells by trapping forces increases as well. This can be seen at higher (several milliwatts) of power. During a scan, cells and smaller pieces of debris can be seen "sweeping" past the field of view in the direction of the laser raster. Trapping effects can be avoided by lowering the laser power, which will likely result in less photobleaching and a better signal-to-noise ratio as well.

4.9   IMAGE SATURATION AND INTENSITY

It is often tempting to increase the exposure time, excitation intensity, or both until the image is "bright enough" to the eye (or computer screen). It is critical, however, to ensure that the signal is in the linear range of the camera (or photomultiplier (PMT) detectors for confocal microscopy). The dynamic range of the detector is also important; however, once again it is important to consider the Nyquist theorem to ensure adequate sampling. An 8-bit detector will provide 256 different gray (intensity) levels. Color, 24-bit cameras actually provide three different 8-bit signals, one for each color (red, green, and blue). The move to a higher-end camera, with 12- or 16-bit grayscale, will provide higher intensity resolution. If one is imaging cells for DNA content, for example, there is a linear difference in haploid, diploid, and aneuploid cells. In this case a lower-bit camera will suffice. However, most 12- and 16-bit cameras also have other features, such as Peltier cooling for lower noise and more sophisticated software control. For more sensitive applications, higher-bit cameras offer better intensity resolution, as the smallest voltage (or photon count) change scales with bit-number of the camera. For a 12-bit camera, there are 4096 possible values compared to the 256 of the 8-bit camera. If the two cameras have the same noise level, then more of the bits in the 8-bit camera are occupied by noise, rather than signal. In most cases, the noise floor of the 12-bit camera is much lower, resulting in even more available signal range. The higher-bit camera can also observe larger differences in brightness, which is necessary for immunofluorescence imaging.

The problem with adjusting the imaging until it "looks right" to the user is that the human eye is a poor gauge for intensity. It is important, instead, to ensure that the brightest signal is below the saturation point of the camera. For a 16-bit camera, the brightest value is 65 536. After adjusting the camera and excitation exposures, an image should be taken and analyzed immediately to ensure that the brightest points do not reach 65 536 counts. If this is the case, then, in reality, the intensity could be much higher, but the camera has saturated in intensity. In many software packages, such as the free ImageJ platform, checking the intensity of the bright regions takes seconds to do. If the image is saturated, decrease the excitation power or the camera exposure time.

The optical microscope is, unfortunately, a subjective instrument. Unbiased operation is difficult to achieve, particularly for fluorescence measurements. There is an inherent tendency to adjust parameters until a visually appealing image appears on screen. This tendency results in overexposed images, excessive photobleaching of the fluorophore, and sample heating. A simple solution to remove some of this subjectivity is to use a brighter monitor. The brighter monitor is well worth the investment, as the image on screen appeases the internal desire for an appealing image at lower excitation power and detector exposure.

4.10 ATOMIC FORCE AND ENVIRONMENTAL SCANNING ELECTRON MICROSCOPY

When one considers the maximum resolution obtained by optical instruments, there are a host of cellular analyses that cannot be performed with sufficient resolution. Below 100–200 nm resolution, even the best optical microscopes fail to resolve many features. This is especially troublesome for bacteria and virus particles. AFM and ESEM measurements are relatively "blind" to chemical and biological information. However, force measurements (for AFM), and immunogold labeling (for ESEM) allow a greater degree of information content. In the AFM case, access to the sample is always required. As a necessity, the cells are left exposed to the laboratory environment. It is therefore critical that experiments be conducted without delay, as viability can decrease rapidly [21]. Sterility is no longer an issue, as bacteria from the lab air will certainly be introduced to the sample within a matter of seconds. Therefore, the more rapid the analysis, the less chance for artifacts from contamination and viability loss.

Anchoring of the cell is particularly important for force measurements, where cells are compressed or sheared by the cantilever. In these cases, it is also convenient and often necessary to include a viability stain. Since many cell-based AFM systems are combined with inverted microscopes, viability of compressed cells can be ascertained. In fact, knowing the viability before compression, one can select the exact type of cell to be tested (i.e., healthy or dead). For ESEM imaging, the cell will enter a partial vacuum and be exposed to the electron beam. Therefore, viability and sterility during analysis are not issues. Cells can be anchored onto any surface that is compatible with the ESEM chamber. The morphology of the ESEM cells should be compared to light microscopy images of the same cell samples, to ensure that even the relatively gentle vacuum of the imaging chamber does not deform cells significantly.

4.11   CONCLUSION

Cell analysis by microscopic methods is the oldest and perhaps most powerful method for elucidating cell structure and function. While other established methods, such as flow cytometry, offer higher counts and better measurement statistics, microscopic imaging can provide detail on the spatial and temporal timescale of the cell. Combined with fluorescence techniques, almost every detail and aspect of the cell can be probed by microscopy. For higher resolution, or newer methods based on mechanical analysis, AFM and ESEM can compliment optical techniques.

Despite the many advantages to cell microscopy, there are many pitfalls as well. Many of these obstacles arise from taking certain aspects of the optical process for granted (Table 4.1). Overlong exposures, irradiances that exceed the saturation point, and phototoxicity all reduce image quality and – in some cases – cell viability. Bleed-through of fluorescence channels, spatial under sampling, and background fluorescence can all obscure the most sensitive analytical signals. This chapter has outlined many of the techniques available for cell microscopy, as well as many practical solutions to culturing cells on the microscope and obtaining the best possible image. Many of the techniques discussed in this chapter, such as staining/fixing protocols, can be used in flow cytometry or other techniques. When combined with the protocols and probes listed in Chapter 9, the optical microscope can be used to elucidate most cell processes with relative ease and with high information content.

REFERENCES

1. Byassee, T.A., Chan, W.C.W., and Nie, S. (2000) Probing single molecules in single living cells. *Analytical Chemistry*, **72**, 5606–5611.
2. Kohl, T., Haustein, E., and Schwille, P. (2004) Determining protease activity *in vivo* by fluorescence cross-correlation analysis. *Biophysical Journal*, **89**, 2770–2782.
3. Zhang, J., Fu, Y., Liang, D. *et al.* Fluorescent avidin-bound silver particle: a strategy for single target molecule detection on a cell membrane. *Analytical Chemistry*, **81**, 883–889.
4. Nguyen, Q.-T., Callamaras, N., Hsieh, C., and Parker, I. (2001) Construction of a two-photon microscope for video-rate Ca^{2+} imaging. *Cell Calcium*, **30**, 383–393.
5. Callamaras, N. and Parker, I. (1999) Construction of a confocal microscope for real-time X-Y and X-Z imaging. *Cell Calcium*, **26**, 271–279.
6. Liu, K., Dang, D., Harrington, T. *et al.* (2008) Cell culture chip with low-shear mass transport. *Langmuir*, **24**, 5955–5960.
7. Liu, K., Tian, Y., Pitchimani, R. *et al.* (2009) Characterization of PDMS-modified glass from cast-and-peel fabrication. *Talanta*, **79**, 333–338.
8. Howell, J.L. and Truant, R. (2002) Live-cell nucleocytoplasmic protein shuttle asay utilizing laser confocal microscopy and FRAP. *Biotechniques*, **32**, 80–87.
9. Hwang, E.Y., Pappas, D., Jeevarajan, A.S., and Anderson, M.M. (2004) Evaluation of the paratrend multi-parameter sensor for potential utilization in long-duration auto-mated cell culture monitoring. *Biomedical Microdevices*, **6**, 241–249.
10. Barkley, S., Johnson, H., Eisenthal, R., and Hubble, J. (2004) Bubble-induced detachment of affinity-adsorbed erythrocytes. *Journal of Biotechnology Applied Biochemistry*, **40**, 145–149.
11. Brock, R., Hink, M.A., and Jovin, T.M. (1998) Fluorescence correlation microscopy of cells in the presence of autofluorescence. *Biophysical Journal*, **75**, 2547–2557.
12. Kohl, T., Haustein, E., and Schwille, P. (2004) Determine protease activity *in vivo* by fluorescence cross-correlation analysis. *Biophysical Journal*, **89**, 2770–2782.
13. Saito, K., Wada, I., Tamura, M., and Kinjo, M. (2004) Direct detection of caspase-3 activation in single live cells by cross-correlation analysis. *Biochemical and Biophysical Research Communications*, **324**, 849–854.
14. van den Engh, G. and Farmer, C. (1992) Photo-bleaching and photon saturation in flow cytometry. *Cytometry*, **13**, 669–677.
15. Shapiro, H.M. (2003) Practical Flow Cytometry, Wiley-Liss, New York.
16. Pawley, J.B. (1996) Handbook of Biological Confocal Microscopy, Plenum Press, New York.
17. Ham, A., Krott, N., Breibach, I. *et al.* (2002) Efficient transfection method for primary cells. *Tissue Engineering*, **8**, 235–245.
18. Chu, G., Hayakawa, H., and Berg, P. (1987) Electroporation for the efficient transfection of mammalian cells with DNA. *Nucleic Acids Research*, **15**, 1311–1326.

19. Lee, W.G., Bang, H., Yun, H. *et al.* (2008) An impulsive, electropulsation-driven backflow in microchannels during electroporation. *Lab Chip*, **8**, 224–226.

20. Masters, B.R., So, P.T.C., Buehler, C. *et al.* (2004) Mitigating thermal mechanical damage potential during two-photon dermal imaging. *Journal of Biomedical Optics*, **9**, 1265–1270.

21. Lulevich, V., Zink, T., Chen, H.-Y. *et al.* (2006) Cell mechanics using atomic force microscopy-based single-cell compression. *Langmuir*, **22**, 8151–8155.

5

Separating Cells

5.1 INTRODUCTION

Cell separations are both an analytical and enabling technique. That is, the cell analysis can serve as the detection event itself, such as in the separation of organelles [1,2] or bacteria [3–5] by electrophoresis. Cell separations can also be a pre-analysis step to purify a certain sample, or to concentrate cells for microscopy or flow cytometry. Therefore, in any discussion of cell analysis one must consider both the source of the cells and the end result desired. For example, it is often necessary to analyze only one type of cell. If the cell can be analyzed in a complex mixture based on a unique identifying parameter, then separation is not required. The most famous case of this is measuring CD antigen expression by flow cytometry. If one to three antibodies (or scatter) can be used to accurately count a specific phenotype of blood cells, then it is preferable to simply analyze by flow cytometry. The speed will be greater, the cell counts higher, and the measurement statistics more accurate. So why separate cells? The answer is that there are many cases when counting cells is not enough, or when the cell cannot be identified as unique in a complex sample. The most important – and prevalent – example is blood. A mixture of cells, plasma, and nutrients, blood is the most important and most complex cell sample available. It is readily accessed and increasingly important to medical diagnostics. Even in the aforementioned case of blood analysis by flow cytometry, typically at least one separation step is performed. Consider the sample itself. Blood is a complex system that is considered dangerous – and rightly so. In the era of blood-borne pathogen awareness, blood is

Practical Cell Analysis   Dimitri Pappas
© 2010 John Wiley & Sons, Ltd

handled with a level of caution that rivals that of explosives and radio-active materials.

Before a routine analysis by flow cytometry, blood is often treated to a Ficoll–Hypaque gradient centrifugation to isolate peripheral mononuc-lear blood cells (PMBCs). The resulting sample is now relatively free of erythrocytes and granulocytes. Whole blood can also be analyzed by cell lysis to remove erythrocytes, but retain leukocytes (monocytes, granulo-cytes, and leukocytes). Both Ficoll–Hypaque gradients and selective lysis are separation steps capitalizing on intrinsic properties of cells (in this case either cell density or resistance to membrane lysis). Other intrinsic properties include electric properties, size, and shape. Labeled properties include cell surface antigen expression, DNA content, and the presence/absence of a particular analyte in the cell. Both intrinsic and labeled properties can be combined in a given separation, and several separation steps can be combined to increase selectivity or purity.

The common analogy to make for cell separations is to examine the wide field of chemical separations. Molecules are routinely separated by different properties in a variety of formats, including gas or liquid chromatography, electrophoresis, and so on. While many of the figures of merit and underlying principles are the same, the difference between chemical and cellular separations lies in the analyte itself. Molecules are relatively hearty when compared to cells, which must be separated in liquid phase (typically aqueous), at or near atmospheric pressure, and with electric fields and chemical/physical conditions that will not harm the cell. In molecular separations, the molecular structure must be retained in most cases. In cell separations, morphology, viability (for unfixed cells), cell function, and often label binding must be preserved. These limitations impose stricter requirements for cell separations, and often result in difficult analyses. However, using the tools and tips provided in this chapter, even the novice can readily perform cell separations to achieve the desired yield, purity, and end result.

5.2   THE CELL SAMPLE

Knowing the composition and desired purification of the cell sample will dictate the method to chose. Figure 5.1 outlines several cell separation methods based on the sample and the desired separation properties. Cell samples can range from relatively straightforward, such as a cultured sample, to the complex. Complex samples are usually "real world" samples, such as blood, other bodily fluids (saliva, urine, semen, etc.),

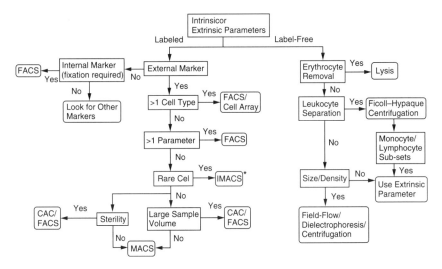

Figure 5.1   Selection of a cell separation technique by cellular parameter. Cell size, internal vs. external properties, and antigen expression are all considered when making a method selection. In addition, sorting by multiple parameters – when possible – improves the purity of the final, sorted samples

or environmental or food samples. In the most difficult cases, the cells are in a matrix that must be removed. A good example of this would be detection of bacteria in food samples. Current techniques for bacterial culture are slow and problematic, as bacteria culled from other sources (such as the operator) may generate inaccurate results. As mention above, blood is a unique sample in its importance, its complexity, and the dangers of testing blood. Knowing what cells are targeted allows one to select the most appropriate separation method. In addition to the sample itself, knowing the desired end result is equally vital to selecting the correct separation method. Will the sample be analyzed after separation? If so, by what method? Preparation for microscopy requires a concentration step if many cells are to be observed in a given field of view. If the samples are to be analyzed by flow cytometry or fluorescence microscopy, additional staining steps may be performed. These steps can be performed before or after the separation, depending on the staining and separation approach used.

When considering the cell sample prior to separation, the matrix, the target cell concentration, and background cell concentration must be considered. Target cells are the cells that are to be isolated with a certain degree of purity and yield. Background cells are the cell types that are not desired, and must be removed. The matrix in which the cells reside places some restriction on the separation methods used. For example, most affinity-based separations of blood require lysis of erythrocytes, as they can interfere with surface interactions (erythrocytes typically outnumber leukocytes by a factor of 1000). Environmental samples, such as pond, river, and lake water, contain a large concentration of solid matter that must be filtered prior to cell enrichment. Even in the case of aqueous cell culture, clusters of cells must be either filtered or disrupted to avoid clogging of the separation device (Table 5.1 lists some of the other common issues when troubleshooting cell separations).

The concentration of the target and background cells also places strict limitations on the sampling method. For example, isolation and separation of rare cells by fluorescence-activated cell sorting (See Chapter 6 for an in-depth discussion of flow-cytometry techniques) is an inefficient use of an expensive (and usually shared) instrument. Tying up an instrument can be avoided if some type of immunomagnetic or affinity enrichment

Table 5.1   Troubleshooting cell separations

Problem	Troubleshooting
Lysis: All cells lysed	Check dilution ratio for lysis agent. Test timing of lysis to ensure target cells remain
Lysis: No cells lysed	Lysis agent may be too dilute. Inadequate lysis time. Cells to be lysed may be resistant to agent, try different lysis agent
MACS: Magnetic beads interfere with subsequent protocols.	Use negative selection (if possible). Remove beads by gently incubation with trypsin-EDTA for several minutes (this may cleave other antigens as well)
MACS: No cells/nonspecific cells	Use compatible magnet
CAC: High nonspecific binding	Use blocking buffer. Adjust flow rate to dislodge background cells while retaining target cells
CAC: Poor retention	Decrease flow rates. Use different antibody clone or capture molecule
FACS: Sorter clogs	See Chapter 6 for discussion of clogs
FACS: Low purity	Re-evaluate sorting parameters. Check sorted sample for types of background cells that may share an antigen with the target
FACS: Low yield	If cell concentration is low, enrich first using MACS. Test sorter with binary cell or bead mixture to ensure sorter is working properly
CE: No cells detected	Electric field is too high (lysis). Adjust to allow separation while maintaining viability
CE: No cell movement	Electric field is too low, or capillary is clogged
Field-Flow: Poor separation	Size/density difference in cells is too low, sort by other methods

occurs prior to FACS sorting. The cell concentration (i.e., cells/volume) is equally important to the background cell ratio. For example, finding a target cell with a concentration of 100 cells ml^{-1} is more difficult if the background (unwanted) cell concentration is 10 000 ml^{-1}. In that case, a 0.1% nonspecific capture rate of background cells would result in 10 background cells captured per milliliter of sample. If the target cell capture rate is 50%, then the total cell number at the end of the separation is 60 for a 1 ml sample. The error, however, is large, as 16% of the cells captured are the background cells. It is possible to then perform a second separation, which would result in 25 target cells and 0 background cells. However, the yield of the target cells has dropped, and any additional side effects of the separation (viability loss, analysis time, etc.) are compounded. The ideal cell separation method, therefore, will have high purity (i.e., percentage of target cells to total retained/sorted cells) with minimal side effects.

The yield of target cells is also important. The recovery efficiency can be thought of as final target cells ml^{-1} compared to the initial cell concentration. It is immediately evident that the cell sample can be diluted or enriched depending on the separation strategy. In flow-based sorting, such as the sorting attachment for Becton–Dickinson's FACSCalibur instrument, the sample is diluted as some amount of sheath flow enters the collection chamber. In other approaches, such as immunomagnetic sorting, the sample can be enriched by resuspension in a smaller volume. However, dilution and enrichment steps aside, yield is also affected greatly by capture efficiency. Using the previous example, a 50% capture rate implies that half of the target cells escape the separation system to waste. Even with 100% purity (i.e., no background cells), a low yield can result in a poor separation. The ideal separation will therefore have a high purity with a high yield.

One must also consider not only the sample type and target cell, but also if the separation will be conducted under sterile conditions. Separations in this case must be conducted in closed systems with sterile reagents, separation media, and so on. It is possible, however, to achieve high cell purity and yield while maintaining sterility. In open-tubular affinity separations of cells, closed fluidics operated in a laminar flow hood minimize the risk of contamination, and allow the captured and recovered cells to be cultured for weeks without contamination [6]. An example protocol for maintaining sterility in a variety of separation steps is discussed later in the chapter.

If the cells are to be cultured or further analyzed, then any label or other alteration to the cell may complicate additional handling steps. In some cases, therefore, it is more advantageous to remove background cells by negative selection (depletion), rather than to isolate the target cells by positive enrichment. For example, if affinity separations are used to isolate a cell sample, there is a possibility that the cell will retain some of the affinity ligands on its surface after separation. In such a case, labeling with a fluorescent antibody after collection may be impaired by steric hindrance. In immunomagnetic separations, bound cells must be separated from the magnetic beads in many cases. It is possible to separate the target from the background (positive) or to isolate the background, leaving a suspension (ideally) containing only the target cells (Figure 5.2). The latter approach is especially useful if the target cell has low affinity for a given separation strategy, but the background cell does not.

In a more complex case, negative selection becomes more cumbersome. Isolating CD14 + monocytes from a PMBC sample, using anti-CD14 columns or beads to isolate cells is relatively straightforward. To achieve the same separation with negative depletion, however, requires several separation steps. One might first deplete B lymphocytes with an anti-CD19 affinity separation. The remaining sample would contain monocytes and

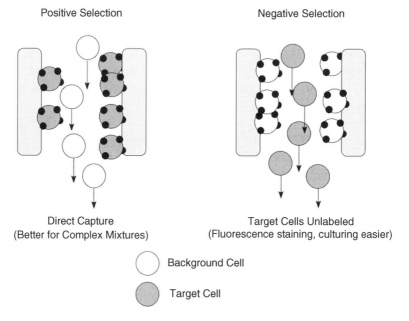

Figure 5.2 Positive and negative selection using MACS methods. In positive selection, the cells of interest are labeled with magnetic particles and captured, while unwanted cells are flushed away. In negative selection, unwanted cells are depleted from the sample by magnetic labeling, while target cells remain in solution. Negative enrichment is useful when the target cell cannot be labeled

T lymphocytes, among others. In one report, a mixture of anti-CD2, CD7, CD16, CD19, and CD56 antibodies were required to negatively deplete the sample and enrich lymphocytes [7]. In this case, the cost and complexity of the separation is much greater than the direct approach. However, many elution methods result in some loss of viability; if the cell sample is particularly sensitive to elution, then the target cell should be negatively selected and passed through the column or detached from the bead unharmed. The background cells can then be removed from the separation medium without regard to viability to regenerate the system.

5.3   LABEL-FREE (INTRINSIC) SEPARATIONS

Of the cell separations performed on a daily basis around the world, the most common is most likely centrifugation. Centrifugation results in density-based isolation of cells, and is particularly useful for blood-cell separations. Centrifugation, also called fractionation, can be conducted in many steps to separate cell fractions. Cell centrifugation can occur in aqueous solution or with the aid of density gradient media. For aqueous separations, centrifugation speed (force) and duration determine the separation type. Separations range from light (1000× gravity) to 100 000× gravity. The extreme separation forces are typically used to separate virus particles and chromosomes from solution. For blood separations, Ficoll–Hypaque medium is typically used to isolate PMBCs. Ficoll–Hypaque is slightly more dense than water, and when blood cells are added to the centrifuge tube (Figure 5.3), the more dense granulocytes and erythrocytes centrifuge to the bottom, while lymphocytes and most monocytes are retained in a thin layer, which can be harvested for additional analysis. Ficoll–Hypaque separations are used with uncoagulated blood and can produce high purity and yield. However, the Ficoll–Hypaque technique can generate sample artifacts [8], such as the loss of certain cell types. It is best, therefore, to compare cell counts of a particular phenotype from lysed blood samples as well as the samples separated by Ficoll–Hypaque.

For aqueous separations, it is possible to conduct multiple separations to separate multiple sample components. However, without the aid of a density-gradient medium such as Ficoll–Hypaque, it is difficult to separate two similar cell types strictly on density. Another difficult – but often attempted – approach is to separate cell debris from the cells by slow centrifugation (<1000×, <5 minutes, depending on centrifuge). Such separations do not consistently result in a debris-free pellet. For example,

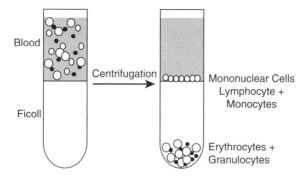

Figure 5.3 Density-based centrifugation using Ficoll–Hypaque techniques. The Ficoll phase is denser than water, and monocytes and lymphocytes remain in the aqueous (blood) phase while denser cells sediment to the Ficoll layer. Peripheral blood mononuclear cell isolation is an example of Ficoll–Hypaque separations, but other cell types can be used

after cell lysis, the lysed erythrocytes have a lower density than intact leukocytes. However, slow centrifugations in aqueous medium yield pellets that contain a larger number of erythrocyte "phantoms." For most cell experiments, however, centrifugation is critical for separating the cells from the medium, allowing suspended cells to be washed prior to analysis by flow cytometry or fluorescence microscopy. Centrifugation can also be used to separate sub-cellular components from lysed cells.

An added benefit of cell centrifugation is that the cell concentration can be adjusted relatively easily. For example, cell concentrations can be increased by resuspending the cell sample in a smaller volume than the original sample. A simple equation for determining cell concentration in the final sample after centrifugation is given below:

$$C_{\text{final}} = C_{\text{initial}} \frac{V_{\text{initial}}}{V_{\text{final}}}.$$

In this case the initial concentration (cells ml^{-1}) is determined by a hemacytometer. The initial and final volumes are readily determined if volumetric pipettes or micropipettes are used.

Selective cell lysis is another method that can be used to separate cells. Lysis can be conducted by several methods and products. The most prevalent are ammonium chloride solutions available from most scientific vendors. The ammonium chloride method is used to selectively lyse erythrocytes in blood samples. Like other, commercial lysis buffers, the ammonium chloride method requires an exposure to the lysis buffer,

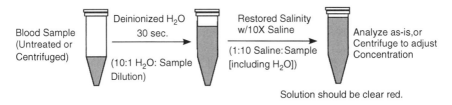

Figure 5.4 Lysis of erythrocytes in blood using a hypotonic solution. A blood sample is exposed to a 10-fold excess of deionized water, during which red blood cells are lysed selectively. Salinity is then restored and additional analyzes can be performed

followed by a wash step to transfer the remaining leukocytes to a non-lysing buffer. Another lysis method that is less expensive and works well for erythrocyte lysis is to centrifuge a blood sample and resuspend the pellet in deionized water for 30 s (Figure 5.4). Alternatively, deionized water can be added directly to the sample in a 10-fold dilution. $10\times$ saline is then added after the 30 s water exposure to restore the solution to an isotonic state [9]. Timing for this salinity approach is important; too short an exposure to the deionized water results in inefficient lysis, and prolonged exposure results in lysis of target cells as well. Cells can then be analyzed in the isotonic solution (i.e., no washing steps), or centrifuged and resuspended in a different solution at a higher or lower concentration. Lysis can also be conducted by automated methods in microfluidic chips (Chapter 7). Cells are moved through the chips and experience either different lysis reagents or changes in salinity. The final cell output is relatively pure leukocytes. As these devices are currently produced in small scale in research laboratories, it is unlikely that they will find widespread use in the near future. These devices do, however, provide continuous introduction of cell sample and lysed cell output.

Cell lysis is also useful for the extraction of nuclei from cells. Nuclei extraction relies on the difference in integrity between the cellular and nuclear membranes. There are many nuclei isolation methods, and most use a lysis method followed by a centrifugation step to isolate the relatively large nuclei from the rest of the cell lysate. The centrifugation step (see Figure 5.5 for a summary) can be conducted in aqueous buffer or in a concentrated (1.8–2.0 M) sucrose solution that cushions the nuclei during separation.

Separation based on cell size requires greater care than selective lysis or centrifugation. Size-based separations have been proposed and demonstrated in the literature for cancer-cell separation, or the more common separation of erythrocytes from leukocytes. In molecular separations,

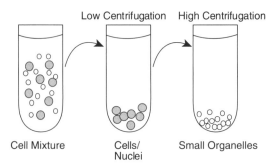

Figure 5.5   Separation of cells and organelles using centrifugation. Lower-g centri-fugation steps remove cells, cell clusters, nuclei, and other relatively large particles. Higher-g centrifugation removes organelles such as mitochondria

size-based separations can occur either via physical interaction, such as size-exclusion chromatography, or by electrophoretic methods. For cells, however, the options are far more limited. Molecular size exclusion, for example, requires a large different (preferably $>5\times$) in molecular weight. With the exception of some of the most disparate cell types, size differ-ences of most cells are small. Examples of instances where an exclusion method might work are erythrocyte/leukocyte separation, or separating mammalian cells from bacteria. In the latter case, removal of bacteria can occur by filtration, although the shear stress on the mammalian cells may result in a large loss of viability. For separation of blood cells, cross-flow microfluidic devices [10,11] and chips using density-based separa-tions [12] have been developed, but have yet to be commercialized.

Another size-based separation method that is worth noting is filtration to remove cell aggregates. This is an often-critical preparation step for flow cytometry, as cell clusters can clog the flow cell or nozzle orifice. Simply passing the cells through the appropriately sized syringe filter (at low flow rate to avoid pressure lysis) can remove unwanted clusters from the sample. As a precaution, cell count and viability (see Chapter 9) should be assessed before and after filtration.

5.4   IMMUNOMAGNETIC SORTING

Magnetic separations, or magnetic cell sorting (MACS) can be used as a preparative step before additional analyses, or as a separation/analysis method by itself. Most magnetic methods take advantage of antibody–antigen interactions between the antibody-conjugated magnetic particles and surface antigens on the cell. It is also possible to buy biotin- or streptavidin-conjugated magnetic particles (for capture of streptavidin- or biotin-conjugated antibodies, respectively), or anti-dye particles. The latter are especially useful for capturing cells using dye-conjugated anti-bodies that can also be used for flow cytometry or microscopy. Several companies currently offer magnetic-based reagents for cell sorting, in-cluding Dynal, Becton-Dickinson, Mitenyi Biotec, and others. The general concept for MACS separations is shown in Figure 5.2. In positive selec-tion, magnetic beads are incubated with the sample for the appropriate time, then the sample tube is placed in a magnetic field and unbound cells are washed away, leaving (ideally) only the target cells in place. After removal from the magnetic field, cells can be washed and analyzed by other methods. Removal of magnetic beads requires disrupting the antigen–antibody bond. Care must be taken to avoid damage to cells during this process (it is best to check viability and cell number with a DNA stain and hemacytometer, respectively). Magnetic beads in this case can be recovered for additional analysis, although this practice is not always acceptable, particularly for clinical samples.

Tip: Trypsin can also be used to separate beads from cells, although some antigens can be cleaved from the membrane in this process, and may have to recover overnight before analysis.

A fast method to test trypsin effects on antigen expression is to test the desired antigen using flow cytometry and fluorescent antibodies. The expression before and after trypsin exposure can be compared and ascertained for each antigen of interest in one afternoon.

Most MACS separations are single particle, single antigen types. For example, extraction of CD34 + hematopoetic stem cells only results in the selection of CD34 + cells and a small percentage of nonspecifically bound background cells. However, it is possible to use particles of different magnetic strength to achieve multi-parameter magnetic sorting. In an example published by Partington et al. [13], smaller magnetic beads are combined with larger ones to achieve two-parameter separations. A stronger magnetic field is required to separate the smaller (50 nm in this case) particles, and this field will also attract the larger (M450 Dynal beads) magnetic beads. Thus, the first selection will separate cells that have either parameter one (small beads), parameter two (large beads), or both. A second separation, using a weaker field that will attract the larger particles but not the 50 nm beads, sorts the two samples.

Manual separations for MACS analyses are the most common, although some flow-through methods, such as the quadrupole magnetic cell sorter [14] and microfluidic methods [15] have been reported. Flow-through methods, like manual MACS separations, can be used in enrichment or depletion mode. Flow-through methods offer distinct advantages over the manual, single-sample method. Flowing a larger volume through such a device results in better enrichment for rare cells. For example, if a cell with 1% occurrence in a 1000 cells ml sample is sorted by a manual (single aliquot) MACS separation with 100% capture efficiency, then for every 1 ml separation 10 cells will be sorted. This example assumes no nonspecific binding of background cells. However, if 10 ml of sample are flown through a device with the same capture efficiency, then 100 cells will be sorted. The error (relative standard deviation, see Chapter 8) in collecting only 10 cells is 31.6%, while the error for 100 collected cells is 10%, and so on. It is possible to create a simple, flow-through device in the laboratory (Figure 5.6). The cell sample can be fed manually, by gravity feed, syringe pump or peristaltic pump, and so on. There is a great deal of flexibility in such a system, but the essential setup in Figure 5.6 will provide fluid transport via tubing past a magnetic field. A waste container collects nonretained cells. To collect magnetically sorted cells, the sample is replaced with a buffer reservoir and any cells suspended in the tubing are flushed to waste. A collection vessel then replaces the waste chamber

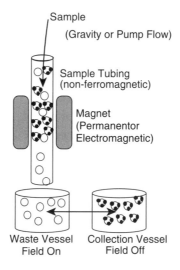

Figure 5.6 Schematic of a flow-through magnetic (MACS) enrichment/separation system. A cell sample is pumped through tubing placed in a magnetic field. In positive selection, waste cells are continuously flushed to waste, while target cells are retained. After the entire cell sample has passed through the device, the magnet is removed or turned off (in the case of electromagnets) and the target cells are collected in a separate vessel. In negative selection, target cells pass into the collection vessel while unwanted cells are retained in the magnetic field

and the magnet is then removed, allowing sorted cells to be collected. Making the tubing as short as practical minimizes dilution during the collection period. It is possible to make such a device using poly(dimethyl-siloxane) (PDMS) microdevices (see Chapter 7), but the complexity is greater and fluid flow rates will be lower. For magnet choice, the magnets supplied by bead manufacturers will result in best compatibility with the magnetic reagent of choice. Strong neodymium and ceramic magnets will also work in most applications, but must be tested for retention efficiency (preferably with a pure cell line). Electromagnets have the benefit of being able to control magnetic strength or to be turned off, but are generally weaker for a given size than permanent magnets.

5.5   CELL-AFFINITY CHROMATOGRAPHY

Affinity separations of cells have been reported in the literature for over 20 years [16]. However, there has been a recent resurgence in the technique, particularly as microfluidic methods have become more commonplace (see Chapter 7). Another cause for the increased attention CAC has received is the expanded need for smaller, less labor-intensive analyzes in resource-poor areas. For example, for CD4 + separations, typically flow cytometry with several labels (at least anti-CD3 and anti-CD4) are used to determine the cell count for AIDS patients. A microfluidic device has been used to separate and count CD4 + cells from whole blood without the need for additional labels or sample-handling steps [17]. As the device is disposable, cross-contamination and blood-borne pathogen exposure are minimized.

Cell-affinity chromatography (CAC) requires a separation volume (a column, microarray plate, or chip) and separation medium. The latter can consist of capture molecules linked to the surface of the separation volume (Figure 5.7) or a 3D medium loaded into the volume, such as a resin or gel [18]. When the capture molecule is coated only on the walls, leaving the majority of the cross-sectional area of the separation volume open, cell interaction is limited to cells that collide with the walls. The collision duration, surface area of contact, and number of molecular bonds formed determine the strength of the retention. Packed volumes – gels, packed bead beds, and so on – offer greater contact area than the open tubular (or open volume) approach, but generally require higher pressure or larger flow rates. In both cases, the prolonged separation or higher pressure may reduce viability significantly. Packed volumes also suffer from reduced elution efficiency, since cells can clog in the different flow paths of the

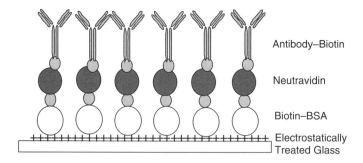

Figure 5.7 An example of separation chemistry for cell affinity chromatography [6,19]. In this case, biotinylated bovine serum albumin (BSA) attaches strongly to a glass surface. Neutravidin [6] or streptavidin [19] is deposited, followed by a biotinylated capture molecule, such as an antibody or aptamer

column or chip. One can think of an affinity separation as a MACS treatment, but instead of beads in a magnetic field, the walls/surfaces of the CAC system serve as the retention platform. Like MACS, CAC methods can select cells in a positive or negative fashion. Unlike MACS, however, cell elution results in a label-free cell without the need for trypsin treatment (see below for elution methods).

When one is considering the type of cell separation to conduct, particularly by affinity chromatography, several parameters must be considered. Does the target cell have a unique antigen, or can a particular capture molecule behave selectively under the right conditions? For example, in the case of CAC separations of CD4 + lymphocytes, one can also retain CD4 + /CD14 + monocytes, leading to a false positive. In flow cytometry, this is generally avoided through forward- and side-scatter measurements, and the addition of a fluorophore-conjugated anti-CD3 antibody (to identify only T cells and reject the monocytes). However, it has been shown that the affinity for CD4 + /CD14 + monocytes is lower when anti-CD4 capture is used (Figure 5.8). One can, therefore, use differences in retention strength to remove cells by shear force [17]. However, if there is a unique antigen for a cell, the separation becomes straightforward. Another parameter to consider is if the separation is to be analytical or preparative. In chromatography terms, analytical separations not only separate compounds, but also provide a measurement (typically concentration or amount) in the process. Preparative separations are generally larger in scale and produce a purified product for additional analysis, culture, and so on. For analytical measurements, one must consider the detection approach. Will on-column detection be preferred? If on-column (or on-chip, etc.) detection is required, then a target cell is immobilized on

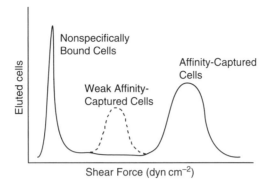

Figure 5.8 Selective elution of cell types in cell-affinity chromatography [6]. Mild washing removes nonspecifically bound cells, while target cells (in this case all CD4 + leukocytes) are removed at higher shear force

the affinity surface and a detection system (primarily optical) counts cells. In this detection scheme, secondary reagents can be added to enhance selectivity. For example, if a cell is captured based on affinity, but only live cells are to be counted, then a viability probe can be added before detection. Secondary reagents, such as additional antibodies for surface antigens other than the target marker, give higher selectivity, as non-specifically bound cells can be rejected based on fluorescence.

The most common example of on-device detection after separation is in the microarray. Microarrays are inherently "multi-analyte" separations, since each array spot isolates a different cell type. On-array detection is almost a necessity, since elution will result in subsequent mixing of the cells that were just separated. Detection can occur by bright-field or dark-field microscopy, or by fluorescence, depending on whether stains/probes are used. On-column detection can be performed, but the limitation in this case is that a 3D volume is imaged, instead of the flat plane of an array. Imaging through a cylindrical (or even square) column requires confocal scanning or image deconvolution to count cells on the walls. For packed-bed separations, on-column detection is impractical in most cases.

Two final considerations for affinity separations are how many cell types will be sorted, and how rare are the target cells? For the former, more than one target cell typically requires an array-type separation. The array approach can also be used to screen multiple capture molecules, such as different antibody clones, for optimal affinity. For array separations, on-chip detection and array scanners allow for multiple phenotypes to be separated. However, separations involving tandem columns [6] can be used to separate more than one cell type as well. Tandem column

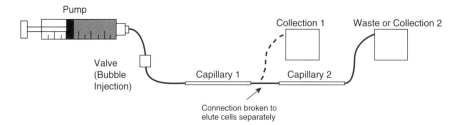

Figure 5.9 Open tubular capillary cell-affinity chromatography. Two capillary affinity columns (using conjugation chemistry from Figure 5.7) are connected in series for multi-parameter cell isolation and recovery. Columns can also be operated in parallel. (Adapted from [6])

separation allows more than one cell type to be separated, as in the array approach; however, the key advantage to multiple-column separations is that each separated cell type can be collected separately. In the tandem case (Figure 5.9), each column can be used to independently separate two cell types, or the first column can be used to remove an interference cell before the second separation. To properly deplete an interference cell (one that shares the target cell antigen used for separation), the first column must be of sufficient length to remove all interference cells before they reach the second column. If the interference cell has a lower affinity than the target, then the depletion column becomes prohibitively long; it is better in this case to use different shear forces to selectively elute each cell type in a single column. Multiple columns can also be used in parallel, where each column separates a different cell type from the same sample independently.

The simple approach shown in Figure 5.9 can be used to create a separation device with minimal cost [6]. The preparation chemistry required is simple, and occurs at room temperature and in aqueous solution [19]. The detailed protocol is given in Chapter 9, but the process is described briefly here. Cleaned glass capillary (inner diameter 50–200 µm) cut to 8 cm in length is used, either as is, or any polyimide coating is burned off to allow better column visualization. Flame removal of the polyimide coating is faster than boiling H_2SO_4 removal, but creates more brittle capillaries, so care must be taken during handling. Successive loading of biotinylated bovine serum albumin, neutravidin, and the biotinylated capture molecule (in this case antibodies) generates the affinity capture surface on the inner walls of the capillary. Teflon tubing of sufficient inner diameter to ensure a tight fit is carefully placed over the capillary. The inlet end is attached to a syringe containing sample (for injection and separation) or buffer (for elution). Unlike most chemical separations, affinity chromatography benefits from continuous injection of the sample. Most chemical separations, based on differential migration due to equilibration between a mobile and stationary phase, require a narrow plug of sample to be injected as rapidly as possible. By continuously flowing cells into the column, rarer cells are enriched, and retained cell number increases. For the open-tubular approach shown in Figure 5.9, the separation ends when the total sample (typically 100–500 µl) is injected, or when the surface of the capillary is saturated with target cells. Cell counting is conducted off column in this case. For validation of phenotype, microscopy or flow cytometry (using fluorophore-conjugated antibodies) is used. Cell purity by this simple, open-tubular separation depends in part on the sample used. Lysed blood is one of the most complex samples available, and purities in excess of 85% for one separation are possible if antibody–antigen affinity is strong and interference cells (sharing the same target antigen) are not present. Purities of cell mixtures in medium approach 95–97%. Multiple separations can increase purity in some cases.

To elute cells by this method, a small bubble (or sequence of 2–5 bubbles) can be injected to dislodge cells [20]. Prior to bubble elution, a gentle washing step can be added to remove cells that are not specifically bound to the surface. Bubble elution results in nearly 100% recovery of attached cells, but can affect viability in some cell types. Checking the cell sample before and after separation using a viability probe (see Chapter 9) can determine the effect of bubble elution on viability. Shear force (high wash speeds) can also be used to remove cells from the capillary columns. In this case the elution is gentler, but can dilute the sample. However, the samples can be enriched several times in CAC, so the overall impact of dilution may be minimal. The elution method will depend largely on whether final concentration or cell viability (for fragile cells) is important.

When cell rarity is concerned, background adhesion (nonspecific binding) and low cell counts are the main cause of error. However, rare cell separation is one of the instances in which CAC can outperform other techniques. Consider the case where a target cell has a relative concentration of one cell per 10^5 background cells. For a FACS sorter (Section 5.10), the maximum sort speed may be several tens of thousands of cells per second. At a concentration of 10^5 cells per ml, one cell will be isolated/sorted per ml of sample, which could occur every 10 s if 1 ml of sample were processed at a speed of 1×10^4 cells s^{-1}. In *Practical Flow Cytometry* [9], Shapiro gives an excellent example of a rare-cell separation by FACS that would yield fewer than 10 target cells in six hours. For the example outlined above, a sort decision every 10 s by FACS would result in

360 cells per hour. However, the assumption in this simple case is that 100% of target cells are sorted, and that 100% of the sorted cells are target cells. In practice, the efficiency and purity of sorting does not approach 100%. It is safe to assume that fewer than 200 cells will be sorted per hour. In one hour's time, the relative standard deviation of counting 200 cells is 14%. Operating a high-speed FACS sorter for excessive lengths of time for a small number of cells is not an economical use of the instrument. As Shapiro mentions, an immunomagnetic enrichment would decrease FACS sorting times tremendously. The same principle applies to CAC. Because affinity separations are typically flow-through applications, a larger volume of sample can be separated than is typically practical by FACS or immunomagnetic separations. In the same case mentioned above, a concentration of one cell in 10^5 cells ml^{-1} could be separated using an affinity separation column with a reasonable flow rate of 100 ml per hour, resulting in a concentrated cell sample. To handle 100 ml of sample by immunomagentic separation would be time consuming, since an affinity separation can be largely an automated process. A microfluidic chip is likely to have a lower volumetric throughput, so for handling large volumes, a column (or array of columns) is recommended.

5.6   AFFINITY CHEMISTRY CONSIDERATIONS IN CAC AND MACS SEPARATIONS

The type of affinity surface depends on the type of cell surface molecule that is to be separated. By the very nature of the separation, the molecule of interest must be found on the outer cellular membrane. CD (cluster of differentiation) antigens are the most common target for affinity chromatography. However, other molecules of interest include phospholipids such as phosphotidyl serine, which is externalized during apoptosis. There are several important aspects to the separation chemistry in affinity separations, much of which also applies to immunomagnetic separations (from the cell-capture molecule standpoint). The first aspect applies to those that will create their own affinity surfaces. Chapter 9 outlines several protocols for conjugating glass or PDMS surfaces to generate affinity surfaces. When considering the affinity surface, one must consider the desired surface concentration, the orientation/functionality of the capture molecules, and nonspecific binding. Nonspecific binding that occurs between the capture molecule and non-target cells cannot be controlled completely without affecting target cell capture. However, nonspecific binding between the nontarget cells and the affinity surface (but *not* the capture molecule) can be blocked by several protocols listed in Chapter 9. Surface coverage can be controlled to an extent by varying reagent concentrations, but quantification of surface coverage is difficult to judge. Atomic force microscopy and single-molecule detection [21] can be used to assess surface coverage; however, for routine fabrication of affinity surfaces, these analyses can be cumbersome. Protocols in Chapter 9 use relatively concentrated solutions to achieve high surface coverage in most cases.

The affinity between the surface capture molecule and the antigen on the target cell surface is the most critical aspect of the cell affinity separation process. The total affinity between the cell and surface is also affected by the antigen density on the surface. While the latter is largely out of control of the scientist performing the separation, the affinity between target cell and capture molecule can be adjusted. It may be necessary to screen a panel of antibody clones to determine which has the largest affinity for cell capture. Unlike flow cytometry or fluorescence microscopy, the affinity must be strong enough to anchor the cell under shear flow. In some cases, such as when a ligand is used instead of an antibody to capture cells expressing a particular receptor, only one type of capture molecule exists (as opposed to different antibody clones). If affinity yield is too low once optimal surface coverage is achieved, then a different method (FACS or immunomagnetic separations) must be pursued.

The final consideration for the affinity chemistry is one of surface interaction. Simply put, if the cell does not collide (gently) with the surface, no affinity capture can take place. In open-tubular separations, laminar flow is readily established in the narrow capillaries or microchips typically used. In laminar flow, a gradient of velocities, with the maximum in the center, is established. One benefit of laminar flow is that cells near the column walls travel slower than those in the center, and thus experience a longer interaction time with the surface. The main drawback of laminar flow, however, is that the flow profile can be considered to be made up of coaxial sheets (laminae) that do not mix laterally (i.e., orthogonal to the direction of flow). Thus a cell in the center of the column is likely to remain there for prolonged periods of time, and the overall surface interaction is low. To eliminate this problem, a narrow column could be used, provided that the inner diameter was not too small to promote frequent clogs. Other way to circumvent the laminar flow limitation is to induce turbulent flow. In order to quickly estimate if laminar flow is established in a column, one can use the Reynolds number, Re. The Reynolds number for open-tubular columns/chips is given by the following [22]:

$$\mathrm{Re} = \frac{d_{c}\rho\langle v\rangle}{\eta},$$

where d_{c} is the column diameter, ρ is the density of the carrier fluid, $\langle v\rangle$ is the average flow velocity, and η is the fluid viscosity. If the Reynolds number is less than 2000, then laminar flow prevails in the column. For packed beds, the term d_{c} is replaced with d_{p}, the particle diameter. For packed systems, Reynolds numbers of 1–100 are sufficient for turbulence [22]. For an open tubular column of 150 μm inner diameter, a volumetric flow rate of $1\,\mathrm{ml\,h^{-1}}$ produces an average linear velocity of about $1.6\,\mathrm{cm\,s^{-1}}$. Using the viscosity and density of water as an example, the Reynolds number is approximately 12. Under open-tubular conditions, this value of Re is far below the criterion for turbulent flow. It is also easy to quickly determine that generating turbulent flow in an open-tubular column with a diameter similar to the one described above with result in prohibitively high flow rates that would likely kill the cells.

A more reasonable method for increasing turbulence is to introduce some type of foreign particles to disrupt flow. Using a low concentration of beads, one can introduce turbulence without resorting to a completely packed bed. Beads have been placed in separation media as the affinity surface [23], but they can also be used at lower concentrations to disturb the laminar flow profile. The beads can be attached using the conjugation chemistry described in Chapter 9. For example, if a streptavidin-coated surface is prepared, then a low concentration of biotinylated beads can be introduced and allowed to attach to the walls. A suitable blocking agent could then be added to ensure that the beads do not capture cells (unless this is desired). This approach is an intermediate one, where just enough beads are added to induce some convection and vortex flow, but the column remains largely open tubular.

In microfluidic formats, there is considerably more flexibility for creating affinity surfaces and optimal flow conditions [24]. Pillars, posts, and other microscopic structures can be added to microfluidic devices to ensure proper flow and surface interactions. The purpose of these posts and other structures is twofold. In some cases, they are merely used to generate a turbulent flow pattern or ensure proper mixing. In others, the additional structures are coated with capture molecules and serve as part of the affinity surface.

In affinity separations, the choice of capture molecule depends largely on the target antigen on the cell surface. In some cases, such as the aforementioned Annexin-V capture of apoptotic cells, the target molecule (phosphatidyl serine) cannot effectively be used to generate antibodies due to its small size. In most cases, however, the cell surface markers of interest can be used to generate a capture molecule through antibody or aptamer technology. Two main classes of antibodies exist (polyclonal and monoclonal) and the type used determines performance level. Since polyclonal antibodies are obtained from animals, they are inconsistent from lot to lot, and also exhibit higher nonspecific binding. Polyclonal antibodies can be raised in a variety of animals, from the ubiquitous mouse to the more esoteric chicken. Chicken antibodies are increasing in popularity due to their relatively lower nonspecific binding when capturing mammalian cells. Monoclonal antibodies are produced when a single B cell is fused with a cancer cell, forming a hybridoma. This single cell can be cultured indefinitely, producing a single antibody structure. In practice, several hybridoma lines are generated, and each culture antibody product is tested for the highest specificity.

TIP: To screen antibody capture from several monoclonal lines or polyclonal lots, an array of antibody spots can be created on a glass surface and tested for optimal cell capture. The protocol is listed below.

To perform this screening test, several linkage approaches to anchor the antibody to a glass surface can be used (see Chapter 9 for protocols). Each antibody is spotted to a specific area of the glass surface, with spots for controls (bare glass as well as each step of the antibody anchoring process). A binary mixture of a target cell and a background cell can then be flowed over the glass slide and incubated for 10–20 minutes (or longer if necessary). Either the target or background cell sample should be stained with a fluorophore that will stain all of the cells in the sample, such as a Hoechst-DNA or cell-tracking dye. This step must occur before the samples are mixed to generate the binary mixture. After several gentle washing steps in phosphate-buffered saline supplemented with 3% bovine serum albumin, fluorescence microscopy of the fluorescent and nonfluorescent cells will reveal the yield and purity for each antibody, as well as the degree of nonspecific binding from each antibody and control spot. To test many antibodies, an array spotter and reader, such as that described in the "LD" array [25], could be used.

PROTOCOL 5.1: SCREENING OF ANTIBODY CLONES

1. Clean glass slides by one of the following methods:
   Piranha solution (caution, this solution is hazardous)
   Boiling H_2SO_4 (caution, this solution is hazardous)

Washing with heptane, isopropanol, and methanol/ethanol. Attach antibodies by one of the methods outlined in Chapter 9. Use careful pipetting to produce well-defined antibody zones).

Note: Use an indelible marker on the underside of the slide to mark the antibody zones)

Note: Make sure to incorporate controls for each reagent added, as well as the bare glass.

2. Remove cells from culture, centrifuge at 900–$1400 \times g$ for four minutes. Decant supernatant for each sample, and resuspend in PBS with 3% BSA.

3. Stain either the target or background cell with a cell-permeant, DNA or cell-tracking stain (see Chapter 9 for examples).

4. Measure each cell sample with a hemacytometer to determine concentration.

5. Mix target and background cells, incubate on array slide for 20 minutes (or longer, depending on antibodies used).

Note: Using accurate aliquots of cell suspension ensures that accurate dilutions and final cell concentrations of the cell mixture can be determined.

6. Gently wash 2–4 times with 3% BSA in PBS.

7. Observe cells by fluorescence and bright field microscopy.

White light microscopy: Determines total number of cells.

Fluorescence microscopy: Determines cells labeled with dye.

Note: Acquire several images at different parts of the antibody spot.

This protocol can be used effectively by hand for 10–20 antibody spots. Since the droplets used to spot the array are small, the reagent use is minimal. For a larger number of antibody spots, an array spotter (available in many core proteomics facilities) can be used.

Antibodies have found widespread success in separations, including affinity capture, MACS, and FACS, because they are specific for a given antigen. However, there are several drawbacks of antibodies that even monoclonal antibodies cannot address. For example, antibodies cannot be directly generated against small molecules, although some methods using small molecules attached to albumin have proven successful [26]. Also, antibodies, like most proteins, are not robust enough for long-term storage. Aptamers, based on RNA or DNA oligonucleotides, are found increasingly as separation agents. Generated in an evolutionary process, aptamers bind target molecules based on shape and hydrogen bonding. Aptamers have no molecular-weight limit with regard to what can be used to generate aptamers. Indeed, entire cells have been used to generate aptamer libraries [27–29].

Aptamers are generated through several processes, although currently the most common method is the SELEX process (Systematic Evolution of Ligands through EXponential Enrichment). In the SELEX method, a library of random nucleotide sequences is passed through a column containing the target molecule. Aptamers that bind to the target are retained and eluted as a mixture. Of course, different aptamers will bind to different degrees, so in this first passage a mixture of sequences is then fed into a polymerase chain reaction (PCR) system to amplify the mixture. The separation process is then repeated, and each successive cycle produces the final aptamer conformation that has the highest specificity. Once that sequence is known, aptamers can be made on a larger scale either by PCR or gene sequencers.

Aptamers work essentially by shape recognition; in fact, in a final SELEX round there may be more than one sequence with near equal affinity for the target molecule. Aptamers are generally more robust than their protein counterparts, and there is no low-molecular-weight limit, allowing more molecules to be used as targets for cell selection. Interestingly, it is also possible to use whole cells as ligands for whole-cell SELEX [27–29]. In this approach, the entire cell is used, so more than one marker may be used to generate a unique aptamer or group of aptamers. This approach is especially useful when there is no identifiable marker, and it can yield aptamers that rival the specificity of those generated for a specific molecule. The SELEX process is slightly modified to accommodate the use of cells, and as stated above, PCR can then amplify the final product.

Aptamers can be used with fluorescent labels for FACS analysis, provided the addition of the fluorophore does not affect the aptamer function (this is also true of antibodies, and any conjugation done in the lab should be tested for capture molecule functionality). Likewise, conjugation to magnetic particles [27] allows for MACS sorting using aptamers, and attachment of aptamers to an affinity surface enables CAC separations.

5.7 ELUTION IN CELL-AFFINITY CHROMATOGRAPHY

Aptamers and antibodies form relatively stable bonds with the cell target molecules. These bonds must be disrupted to elute the cell. In array separations, elution must be avoided, but in all other analyses it is desired. As mentioned in the introduction of this chapter, most molecular-affinity chromatography methods to elute bound analytes (pH and/or ionic strength gradients) generally are not amenable to viable cells. There are three main strategies for eluting live (or fixed) cells. First, a gentle chemical gradient [30] can elute cells with high viability. Cells eluted from such gradients must be washed and resuspended in a suitable buffer or medium to avoid adverse effects. The second method is to use a high flow rate and elute the cells by shear force. The benefit of using shear elution is that the cells can be kept in whatever buffer or medium is desired. However, after all of the steps taken to separate and enrich cells, the final elution step by shear flow results in dilution, which in some cases is detrimental to the original purpose of the separation. The third method of cell elution is to introduce a short sequence of bubbles [6,20]. Bubble elution is highly efficient, but should be used with care, as viability can be affected. Bubble elution, or any mechanical disruption of cells, should be characterized with test samples both for long-and short-term effects on viability. A 20-minute stain and test with an impermeant DNA dye (Chapter 9) will determine short-term loss in cell viability. However, when developing a bubble elution protocol for the first time, or with a new cell type, it is advisable to also perform the assay several hours later (4–6 h), as well as the next day. Apoptosis assays using Annexin-V can be combined with an impermeant DNA dye for a more complete picture of the cell sample viability. In some cases, a cell line may appear highly viable (>80%) immediately after elution, only to suffer from significant viability loss over the next 24 hours. If viability is paramount for a sample (i.e., sorting rare cells for re-culture), then it is advisable to spend the upfront time characterizing any affinity separation. If cells are fixed prior or after separation, then viability characterization is unnecessary.

5.8   NONSPECIFIC BINDING IN CELL SEPARATIONS

In traditional analytical measurement, the limit of detection is defined as a signal (or concentration) that is some interval larger than the mean of the blank signal (or concentration). The International Union of Pure and Applied Chemistry (IUPAC) definition of the limit of detection for chemical analytes is three times the standard deviation of the blank signal, divided by the measurement sensitivity. In practice, this definition results in inaccurately low limits of detection, since most scientists ignore the stipulations of the IUPAC limit. In a landmark paper by Winefordner and Long [31], many limits of detection were found to be falsely low, as the

IUPAC definition requires both the intercept and slope error of a calibration curve to be much smaller than the error (noise) in the blank. The limit of detection – whether the IUPAC or the more refined and robust propagation of errors approach by Winefordner and Long – can be applied to a cell separation. The cell limit of detection (N_L), or capture, can be determined as follows:

$$N_L = 3s_B = 3\sqrt{n_B},$$

where s_B is the error (uncertainty) of a blank sample, and can be represented by the square root of the number of captured or measured background cells (n_B). A "blank" is difficult to perform in cell separations, especially when the sample is complex, such as blood. If one wanted to separate CD34 + stem cells from blood, a blank would then have to contain every component of blood except the CD34 + stem cells. To produce this in a realistic setting, blood would have to be depleted of the target cell, which is the exact aim of the separation experiment in the first place.

More often, nonspecific binding is used to determine cell detection limits. Nonspecific binding of background cells is the main source of uncertainty in any cell analysis, especially as the relative target cell concentration decreases. If a separation has a 1% nonspecific binding (i.e., 1% of nonspecific cells are captured alongside the target cells), then a sample with a target cell concentration of 1% would result in a captured population that is 50% target cell and 50% background cell. The separation becomes more problematic as the target cell concentration approaches the nonspecific binding value. This example assumes a set volume of sample has been introduced into the separation medium. If the sample is continuously flowed through the separation medium, then the target cell may be enriched faster than the degree of nonspecific binding.

To verify that a target cell is present, only by cell count, one must prove that the captured cell number is statistically greater than the nonspecific binding. Using the limit of detection defined above, the captured cell number must be three times greater than the square root of the background cell number captured during the separation. If 300 μl of sample (with a 10^5 total cells ml concentration) has a 1% nonspecific binding, then 300 cells would be captured in the separation column or channel. The target cell limit of detection would then be 352 target cells (i.e., s_B is 17.3, resulting in $3s_B = 52$). The total cell number in the separation would be 652 (target and background cells). The lower the nonspecific binding, the better the limit of detection. The limit of detection, however, is the lowest number of cells that can be *reliably* detected as different from the background cell count. The limit of identification is twice the limit of detection (requiring 404 target cells to reliably identify the sample). The limit of quantification is $10s_B$, requiring 573 target cells to be detected before quantification could be considered reliable. These guidelines are for chemical concentrations, and while they can be translated to cell samples, they must be used with caution, as other factors may affect analytical performance. For example, co-capture of two different cell types that share an antigen (when only one cell is the target cell) would not obey the limits of detection that are so routinely used in chemical analysis. These limits, however, are useful in predicting the success of a rare cell separation, as nonspecific binding will mask any retained target cells.

Nonspecific binding is affected by several factors, including interactions between the cell and the untreated surface such as glass or magnetic particles, the cell and the immobilization chemistry (e.g., exposed neutravidin), and the cell and the capture molecule itself. In the last case, nontarget cells can interact with the capture molecule, particularly when antibodies are used. There are several steps that can be implemented to reduce nonspecific binding, according to the type of interactions.

1. **Surface Interactions.** If the substrate is glass, adding 3% albumin to the separation buffer can reduce nonspecific binding. As an alternative, commercially available blocking buffers can be used. Blocking buffers are proprietary mixtures and are made of protein or protein-free mixtures, depending on the application. Some empirical work will have to be performed to ascertain which blocking buffer will work best. Typically, the blocking buffer is added *after* all of the conjugation steps are completed. For PDMS and other hydrophobic polymers, coating with a blocking buffer, copolymer additives [32], or albumin will reduce nonspecific binding to a degree. However, the hydrophobic nature of PDMS results in adhesion of many lipids and other nonpolar molecules. For bacterial separations, the addition of dilute cranberry juice [33] can reduce nonspecific binding of many species.

2. **Immobilization Chemistry.** During the conjugation process, successive layers of molecules are conjugated to create the final affinity surface. During each step, incomplete surface coverage can result. For example, if avidin, streptavidin, or neutravidin are immobilized on a surface (for use with biotinylated capture molecules), endogenous biotin or nonspecific interactions with the cell surface can result in capture. Some level of blocking, using a biotinylated

molecule that does not interact with the cell, should be implemented after the biotinylated antibody is added (Chapter 9). In general, adjusting reagent concentrations to ensure adequate surface coverage will minimize this type of nonspecific binding.

3. **The Capture Molecule.** Careful attempts to reduce nonspecific binding can be ruined if the capture molecule itself displays significant nonspecific binding. Fc receptors on cell surfaces can interact with antibodies regardless of the antigen type; blocking using serum can reduce this effect. Also, moving backward on the evolutionary ladder can also reduce nonspecific binding of the capture antibody. For aptamers, nonspecific binding may indicate that the SELEX process must be repeated to obtain a more specific aptamer, or that nonspecific binding is occurring from something other than the capture molecule (See items 1 and 2, above).

5.9   SEPARATION OF RARE CELLS

When the target cell concentration drops below approximately 0.01% of the total population, the separation becomes more difficult. The large disparity in relative target cell concentration is compounded both by nonspecific binding and the need to identify a unique marker or parameter with which to separate the cells. Of the many applications of rare cell separation, perhaps the most commonly attempted is the isolation of circulating tumor cells (CTCs) in blood. The necessity for this type of separation is evident, as early detection of tumor metastasis could improve the prognosis of recovery for cancer patients. In the best cases on nonspecific binding, 0.01–0.1% of background cells are captured. For most rare cell separations, nonspecific binding makes selective cell capture difficult. Even in FACS sorting of rare cells (Section 5.10), multiple passes are sometimes required if the rare cell purity is too low. In the cases of more abundant rare cells (e.g., 0.01% relative sample concentration) and the best nonspecific capture (0.01%), the resultant sample is only 50% pure. For DNA testing or reculture of the rare cell, sample purity should approach 100% (particularly for reculture of separated cells).

The separation of rare cells is made more difficult when one considers that one parameter alone is not sufficient for rare-cell isolation and sorting. In *Practical Flow Cytometry* [6], Howard Shapiro makes a convincing case that for rare-cell sorting, more than one parameter (and often more than two) is needed to accurately sort cells. The same can be applied to any cell separation type. The main problem with non-FACS

Table 5.2   Tandem cell separation techniques

Technique	Applications
MACS	Preconcentrate for flow sorting (single aliquot)
MACS	Continuous-flow preconcentrator for flow sorting
MACS	Negative selection to deplete background cells from CAC, FACS, or other separation
CAC	Can use multiple CAC columns–chips in series or parallel
CAC	Continuous-flow preconcentrator for flow sorting
Field-Flow Fractionation	Continuous separate before CAC, MACS, FACS

separations is that most take advantage of one cell property or cell surface marker. In the case of CTCs, one could separate by size (on the assumption that most CTCs are larger than normal leukocytes in blood), followed by a specific marker for that cancer cell type. Even in this case, the size separation likely requires larger size differences than those between CTCs and leukocytes to separate with sufficient purity. Also, the likelihood that the rare cell in question will have two unique cell surface markers (so that the cell can be separated by MACS or affinity methods) is low. More likely, two different methods must be used. Table 5.2 lists several different separation types that can be used in tandem when rare cell separation is required. The use of different methods – such as MACS, to enrich a rare sample before FACS sorting – improves separation in the final, separated sample.

5.10   FLUORESCENCE-ACTIVATED CELL SORTING

It might, at first glance, seem more appropriate to include any discussion of FACS analysis with Chapter 6 (Flow Cytometry). However, in this discussion, the mechanisms of sample flow, detection, event counting, and so on, are kept to a minimum. Since FACS is primarily a separation method, it will be treated in this chapter. Those interested in the fundamental operations of a flow instrument should read Chapter 6, and also seek out several excellent texts on the matter (e.g., [6]).

The key benefit to FACS analysis is that, unlike affinity or MACS separations, the key parameter for sorting does not have to be an extracellular antigen. It is, of course, common to separate cells by FACS using a fluorescently labeled antibody for a cell marker, but other modes of operation exist as well. For example, staining cells with Hoechst dyes allows cells to be sorted according to their place in the cell cycle (Chapter 1, Figure 1.6). Other examples of separations that can essentially be done only by FACS include intracellular antigens and separation based on light scatter. FACS sorting can be conducted using large-scale, high-speed sorters or more compact, microfluidic based systems. Like MACS and CAC separations, FACS sorting can be conducted under sterile conditions, and can also be used to enrich cells. Unlike MACS and CAC, multiple parameters can be used simultaneously to separate cells. For example, staining for CD71 (human transferrin receptor) and DNA content can be used to isolate senescent cells from proliferating ones. In this case, cells in the G0/G1 cell cycle phase that stain CD71-negative would be sorted, while proliferating cells pass to waste.

FACS sorters, as the name implies, primarily use fluorescence as the readout mechanism prior to the sort decision. There are three steps in a sorting event, all of which occur at high speed. The readout of fluorescence and light scatter identify the cell. A sort decision, based on logical gates on the fluorescence data, then triggers the sorting mechanism. The readout and decision do not change much across instruments, in that fluorescence intensity, polarization, or wavelength are used to trigger an event. The decision can be comparative (greater or less than a threshold value) or can occur when a value falls within a certain gate (or several gates, if more than one parameter is used). These gates or thresholds are set based on control samples prior to the sorting of the analytical sample. Setting gates for FACS analysis requires positive and negatively stained controls. Using the aforementioned CD71/cell cycle example, three cell controls would be needed: a blank sample (no reagents), anti-CD71 only, Hoechst only, followed by samples containing both anti-CD71 and Hoechst stain.

The actual sorting, unlike MACS and CAC, is a mechanical process. In the sorting event a cell is diverted from an original trajectory to a new one. Figure 5.10 shows the process for a droplet-based sorter, the oldest and most common mechanism. Cells exit a flow nozzle (see Chapter 6 for a discussion on the flow cytometer flow system) as a stream in air, where a laser (or group of lasers) irradiates the cell and fluorescence is measured. A piezo transducer on the flow nozzle induces a uniform droplet formation as the stream breaks apart. Cells are encapsulated in a droplet during this period, and pass with a steady velocity through a pair of plates. When a sort decision is made, a delay is triggered so that the cell measured by the laser(s) passes through the charging plates and the droplet is charged. A pair of electrostatic deflection plates deflects the charged droplets, while uncharged droplets pass through to a collection or waste vessel. FACS instruments operate by charging the target cells and passing all others to waste, or by charging all waste cells (the latter is advantageous if droplet charging affects viability or cell function).

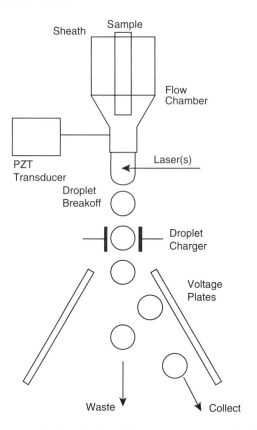

Figure 5.10   Schematic of a droplet-based fluorescence-activated cell sorter. Droplets formed from the flow-chamber nozzle are detected by a laser for the desired cell fluorescence. Droplets containing a target cell are given an electrostatic charge and then deflected to a collection well by two plates. Droplet sorters can also charge droplets not containing target cells, if electrostatic charging of the drops affects cell viability

While droplet-based sorters are the fastest and most common FACS instruments, there are several other approaches to cell sorting in a fluid stream. Partec sorters use a piezo-driven valve to divert the sample stream and collect target cells. The technique is not as fast as droplet sorters, but is a closed system (advantageous for biohazardous materials). The FACSCalibur sorter from Becton-Dickinson collects cells by moving a collection tube into the sample stream. Microfluidic sorters have been developed [34,35], and almost all are built in the lab and integrated onto microscopes. The sorters can be made with integrated pumps and valves [35], but are generally slow ($<1000\,\text{cells}\,\text{s}^{-1}$) and are not currently available commercially. Sorting generally requires higher speed to minimize shock to cells. This is particularly true when sorting primary cells from tissue extracts, as the time between extraction, sorting, and culture can diminish cell yield.

When sorting cells, regardless of the final application, one must consider the effects of the sort itself on cell health. For fixed cells, the criteria for successful separation are less stringent, as viability, function, and sterility are not considered. For live cells, however, the effect of droplet sorting can lower viability. To test for possible negative effects during FACS separation, a cell sample (either pure or a mixture) can be sorted, and viability compared to a control. It would also be better to conduct apoptosis assays at hourly intervals for several hours, to assure that long-term effects on cell function are not introduced by the sorting event. For droplet sorting, for example, the electric field on the droplet may lyse cells, and the collection of droplets may result in mechanical stress on the cells. For fluidic (closed system) sorters, changes in shear force and pressure may affect cells.

In some cases, a target cell with initial concentration in the sorter may be diluted. This is especially problematic in fluidic sorters that have a portion of the sample or sheath fluid enter the collection channels continuously. In these cases, an enrichment step must be conducted, unless the initial cell concentration was sufficiently high that dilution is not a problem. Another consideration, regardless of the sorting mechanism, is the issue of sterility. Modern stream-in-air (droplet) sorters have built-in measures to reduce aerosol exposure, which is critical for reduction of operator risk. However, the cell samples themselves may require sterile sorting for some experiments. For example, purification of primary cells for culture requires sterile conditions or the introduction of antibiotics.

Sterile sorting requires a closed fluid system, or a stream-in-air system operated in a sterile hood. An example of the latter is the InFlux sorter from Cytopoeia (now part of Becton-Dickinson). The InFlux instrument is a high-speed droplet sorter that is highly customizable, and in one configuration can be built into a laminar flow hood/biosafety cabinet. The flow and sorting section of the instrument are housed in the laminar hood, while the lasers, detectors, and electronics are located on the backside of the hood. The benefit of this approach is that the components that do not need to be sterile do not take up precious hood space, while the fluidic and sorting sections are maintained in a sterile environment. It would be conceivable that other sorting instruments could be built into a laminar flow hood. The InFlux has also been operated with a cartridge system that allows a reel of bubble chambers (formed in plastic) to collect individual sorted droplets, which are then heat-sealed to form a sterile container for individual bacteria.

It is impractical at times to modify both sorter and laminar hood. In these cases, a closed fluidic sorter will be a better option. Passing a sterilizing agent (that is compatible with the sorter) through the fluidics will sterilize the system. A sterile sample can then be sorted and collected without contamination from bacteria or other cell lines. It is important to note that the original sample must be sterile, otherwise the steps taken to sort under sterile conditions are wasted. The sample can also be sorted using a droplet or fluidic sorted in nonsterile conditions, and subsequently treated to remove bacteria and fungi. Using this approach, the accidental introduction of bacterial and fungal species can be mitigated. However, in some cases the use of antibiotics may alter cell function (Chapter 3). As a general rule, if the sample must be sterile after sorting, the best approach is to sort the sample under sterile conditions.

5.11   SORTING PARAMETERS

Sorting, as mentioned above, is a logic-driven process, where a fluorescence signal(s) is compared to those of reference or control samples. Fluorescence parameters used to make sorting decisions include fluorescence intensity (height and/or area) of cell surface antigens, intracellular species, DNA content, light scatter, and uptake of certain dyes. One of the main benefits of FACS separation is that autofluorescence can be used to isolate cells. For example, rod-shaped myocytes, isolated from rat heart ventricles, were sorted using three parameters [36]. Cell size (forward and side scatter), autofluorescence, and time-of-flight were all used to set the sorting decision. The third parameter was important to sorting cells of rod-type dimensions, as the cell length was longer than that of other cell types. It is therefore important to know what type of cell sample is to be isolated. Knowledge of the cell sample in this case allowed sorting without additional reagents. This example showcases the instances where FACS sorting is the only available option. Despite the rapid growth in cell separation methods – many of which are too new to discuss in a practical text – FACS separations will continue to play an important role in cellular analysis.

5.12   OTHER SEPARATION TECHNIQUES AND CONSIDERATIONS

Fixing cells is at times necessary (to preserve structure or chemical content) or desired (to reduce biohazard risk), and fixed cells require less strict separation protocols. Fixed cell sterility is not an issue, nor is the need to maintain cell viability. For other cases where cells are not fixed, necrosis and apoptosis can impact cell yields and effect subsequent applications. There are several factors that can lower viability and/or induce apoptosis in the sample. First, the effect of shear force on the cells can result in cell death. Shear can be introduced at several points in the experiment. For example, injecting samples into a separation device using small bore needles and tubing may yield a large shear stress. Shear stress in a cylindrical channel is given by:

$$\tau = \frac{4\mu Q}{\pi r^3}.$$

Where Q is the volumetric flow rate, μ is the fluid viscosity, and r is the channel radius. It is clear that as the radius of the channel decreases, the shear force increases. To circumvent this effect, cells should be introduced in the widest lumen that is practical. If too large a syringe/tubing/channel is used, other flow effects may alter the separation outcome. Shear in the separation itself is largely unavoidable, since most separation methods require some type of fluid medium to move target and background cells away from each other.

In glass channels, capillaries, and chips, there is an added complication since the glass is largely not permeable to gases. In many cases, the flowing buffer or medium in separation will permit fresh nutrients and oxygen to reach the cells. However, if flow rates are low (to avoid shear), or if stop-flow methods are used, then gas exchange becomes critical. It should be noted that stop-flow methods are often used when a cell-capture molecule affinity is too low to form sufficient bonds in the presence of flow. To impart some level of gas exchange, open channels can be used, or a gas-permeable membrane can be introduced. These techniques are discussed in greater detail in Chapter 7, and can be implemented in separation devices to prolong cell viability. Another, often overlooked, factor in viability during separation is temperature. Cells will survive at reduced temperatures (4–20 °C), and the survival time depends largely on cell type as well as other environmental factors. Incubating the separation medium (see Chapter 4 for examples of stage heaters) can improve viability for long-term cell separations.

Other techniques that are available, but not widespread at this time, are field techniques such as field-flow fractionation, electrophoresis, and dielectrophoresis. Field-flow techniques can be assembled in the laboratory without microfluidic fabrication needs [37] and take advantage of

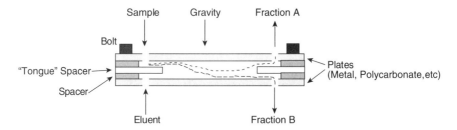

Figure 5.11   Field-flow fractionation. Cells separate between fluid laminae based on density and size. Field-flow fractionation is a binary cell separation, and is operated in a flow through manner, allowing for high cell yield

differential sedimentation in laminar flow. The transport velocity (U) across laminae is expressed as the sedimentation force divided by the friction coefficient f as follows:

$$U = \frac{\Delta\rho G d^2}{18\eta(f/f_0)} = sG.$$

In this case, $\Delta\rho$ is the difference in cell and fluid density, G is the acceleration due to gravity, d is the cell diameter, η is the fluid viscosity, and f/f_0 is a correction factor for cell deviation from a perfect sphere. In field-flow cell separations, cells of different density, size, and shape can be exploited to separate cell types. Figure 5.11 shows a simple device for field-flow cell separations that can be constructed with the aid of a machine shop [38]. Similar devices are available commercially from Postnova Analytics.

Dielectrophoretic methods use electric fields (generated by surfaces patterned with electrodes) to separate cells based on differences in cell size and dielectric properties. The cell diameter and membrane capacitance determine cell migration in an electric field. The patterned devices are not simple to construct, and are not widely available due to the specialized equipment required to fabricate them. However, dielectrophoresis is also being integrated with field-flow techniques to improve separations through the additional of multiple, orthogonal parameters [39]. By combining dielectrophoretic and field-flow methods, cell size, shape, density, and capacitance can be used simultaneously for sorting and isolation.

Capillary or microfluidic electrophoresis is also an interesting cell separation approach, in that it is label-free and also easy to implement

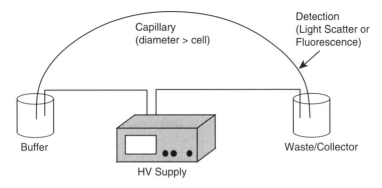

Figure 5.12   A conventional capillary electrophoresis system, used for bacterial cell separation. Current CE systems, using laser-induced fluorescence or light scatter, can be directly used with bacteria. Microfluidic electrophoresis systems can also separate bacterial cells (see text)

in most laboratories. An example electrophoresis system is shown in Figure 5.12. Capillary separations occur via electric field under electrophoretic and electro-osmotic flow [22]. The general appeal of capillary electrophoresis (CE) is that it is simple to implement, and has high separation efficiencies. Detection can be based on absorption, fluorescence, electrochemistry, or coupled to mass spectrometry or other techniques. However, the majority of electrophoresis separations are not cellular, but rather chemical. The standard operating conditions are therefore not optimized for cell separations. Separations of sub-cellular components, sometimes referred to as chemical cytometry of single cells, has been developed and is readily implemented [40]. The buffers used in CE are generally of different osmolarity than what is required for cells (Chapter 3), and the electric fields present in the capillary can lyse cells. Cell lysis by electric field is performed routinely in "chemical cytometry" measurements of single cells [41,42].

Detection in CE for chemical analysis varies by technique. However, for cell work, several practical considerations limit the number of detection schemes. Absorption, one of the most common CE detectors, cannot accurately measure the minute concentrations of analytes in cells. In general, light scattering and fluorescence are well suited to cell detection, as they have been applied in flow cytometry for decades. Fluorescence detection (shown in Figure 5.12) has the added benefit of increased information content through the use of probes and labels. For example, DNA probes have been used to assay viability in bacterial and yeast cells during CE separations [43]. Bacterial and yeast cells are less susceptible to damage in CE experiments, and therefore have been reported most frequently in the literature. Detection of bacterial cells by CE is complicated by the motility of bacteria, which leads to zone broadening and reduces separation efficiency. Lantz *et al.* [3] introduced a low-fluorescence blocking agent to reduce cell motility. Using this approach, they were able to separate and detect individual bacterial cells. CE is a "plug" technique, in that a small sample volume is injected into the capillary and separation is performed. Unlike other cell separation approaches, where large numbers of cells are sorted, CE of intact cells can be used to detect and identify certain cells, or identify different cellular conditions. Before separation by CE, cells (bacterial or yeast) of interest should be incubated in the separation buffer (including detection reagents) to determine effects on viability. The osmolarity of the solution, as well as pH, and so on, will have to be optimized to balance between cell viability and separation efficiency.

5.13   CONCLUSION

Cell separations are perhaps one of the most difficult types of cell analysis. As both an analytical and preparative technique, cell separations serve both as an enabling science and a measurement in itself. There are, however, several challenges to successful cell separations. First, knowing the sample will eliminate many artifacts during separation. In addition, an understanding of the parameters used for separation will ensure greater success. An example of failing to heed both of these caveats is the perennially important separation of CD4 + lymphocytes for AIDS analysis. Not knowing the intricacies of the sample (peripheral mononuclear cells, whole or lysed blood) and the capture parameter (CD4 +) could produce separated samples consisting of CD3 + /CD4 + lymphocytes and CD4 + /CD14 + monocytes. This example represents a major hurdle for separations, and one that must be overcome in order to bring cell analyzes to those that need them most.

When separating rare cells, one must remember that any rare cell analysis will be plagued by two factors. First, the rarity of the cell may necessitate large sample volumes and/or long separation times, neither of which may be feasible. Second, nonspecific binding of non-target (background) cells requires that some other parameters also be included to identify cells. This second parameter could result in a second separation, or could be a fluorescence parameter to identify the separated cells as target or background cells. As the number of parameters increase, using a separation for each of them becomes less practical, with the exception of FACS separations, which can measure all desired parameters simultaneously.

Cell separations will increase in use and efficiency as new capture molecules and microfluidic technologies become available. In addition to the widespread use of FACS and MACS, CAC, with its ease of implementation and high volume throughput, will become even more widespread. The topics covered in this chapter should encourage those interested in separating cells to take the small steps required to set up separation systems in their own lab for cell purification, sorting, and analysis.

REFERENCES

1. Navratil, M., Terman, A., and Arriaga, E.A. (2008) Giant mitochondria do not fuse and exchange their contents with normal mitochondria. *Experimental Cell Research*, **314**, 164–172.
2. Feng, J., Xie, H., Meany, D.L. *et al.* (2008) Quantitative proteomic profiling of muscle-an age-dependent protein carbonylation in rat skeletal muscle mitochondria. *Journal of Gerontology: Biological Sciences*, **63**, 1137–1152.
3. Lantz, A.W., Bao, Y., and Armstrong, D.W. (2007) Single-cell detection: test of microbial contamination using capillary electrophoresis. *Analytical Chemistry*, **79**, 1720–1724.
4. Hjerten, S., Elenbring, K., Kilar, F. *et al.* (1987) Carrier-free zone electrophoresis, displacement electrophoresis and isoelectric focusing in a high-performance electrophoresis apparatus. *Journal of Chromatography*, **403**, 47–61.
5. Desai, M.J. and Armstrong, D.W. (2003) Separation, identification, and characterization of microorganisms by capillary electrophoresis. *Microbiology and Molecular Biology Reviews*, **67**, 38–51.
6. Wang, K., Marshall, M.K., Garza, G., and Pappas, D. (2008) Open-tubular capillary cell affinity chromatography: single and tandem blood cell separation. *Analytical Chemistry*, **80**, 2118–2124.
7. Ujam, L.B., Clemmitt, R.H., Clarke, S.A. *et al.* (2003) Isolation of monocytes from human peripheral blood using immuno-affinity expanded bed absorption. *Biotechnology and Bioengineering*, **83**, 554–566.
8. Renzi, P. and Ginns, L.C. (1987) Analysis of T-cell subsets in normal adults. Comparison of whole blood lysis technique to ficoll-hypaque separation by flow cytometry. *Journal of Immunological Methods*, **98**, 53–56.
9. Shapiro, H.M. (2003) *Practical Flow Cytometry*, 4th edn, Wiley-Liss.
10. Murthy, S.K., Sethu, P., Vunjak-Novakovic, G. *et al.* (2006) Size-based microfluidic enrichment of neonatal rat cardiac cell populations. *Biomedical Microdevices*, **8**, 231–237.
11. VanDelinder, V. and Groisman, A. (2007) Perfusion in microfluidic cross-flow: separation of white blood cells from whole blood and exchange of medium in a continuous flow. *Analytical Chemistry*, **79**, 2023–2030.

12. Chen, X., Cui, D., Liu, C., and Chen, J. (2007) Continuous flow microfluidic device for cell separation, cell lysis, and DNA purification. *Analytica Chimica Acta*, **584**, 237–243.

13. Partington, K.M., Jenkinson, E.J., and Anderson, G. (1998) A novel method of cell separation based on dual parameter immunomagnetic cell selection. *Journal of Immunological Methods*, **223**, 195–205.

14. Williams, P.S., Zborowski, M., and Chalmers, J.J. (1999) Flow rate optimization for the quadrupole magnetic cell sorter. *Analytical Chemistry*, **71**, 3799–3807.

15. Xia, N., Hunt, T.P., Mayers, B.T. *et al.* (2006) Combined microfluidic-micromagnetic separation of living cells in continuous flow. *Biomedical Microdevices*, **8**, 299–308.

16. Pappas, D. and Wang, K. (2007) Cell separations: a review of new challenges in analytical chemistry. *Analytica Chimica Acta*, **601**, 26–35.

17. Cheng, X., Irimia, D., Dixon, M. *et al.* (2007) A microfluidic device for practical label-free CD4 + T cell counting of HIV-infected subjects. *Lab on a Chip*, **7**, 170–178.

18. Dainiak, M.B., Kumar, A., Galaev, I.Y., and Mattiasson, B. (2006) Detachment of affinity-captured bioparticles by elastic deformation of a macroporous hydrogel. *Proceedings of the National Academy of Sciences USA*, **103**, 849–854.

19. Jing, R., Bolshakov, V., and Flavell, A.J. (2007) The tagged microarray marker (TAM) method for high-throughput detection of single nucleotide and indel polymorphisms. *Nature Protocols*, **2**, 168–177.

20. Barkley, S., Johnson, H., Eisenthal, R., and Hubble, J. (2004) Bubble-induced detachment of affinity-adsorbed erythrocytes. *Journal of Biotechnology and Applied Biochemistry*, **40**, 145–149.

21. Wayment, J.R. and Harris, J.M. (2006) Controlling binding site densities on glass surfaces. *Analytical Chemistry*, **78**, 7841–7849.

22. Giddings, J.C. (1991) Unified Separation Science, John Wiley and Sons.

23. Pregibon, D.C., Toner, M., and Doyle, P.S. (2006) Magnetically and biologically active bead-patterned hydrogels. *Langmuir*, **22**, 5122–5128.

24. Nagrath, S., Sequist, L.V., Maheswaran, S. *et al.* (2007) Isolation of rare circulating tumour cells in cancer patients by microchip technology. *Nature*, **450**, 1235–1239.

25. Belov, L., Mulligan, S.P., Barber, N. *et al.* (2006) Analysis of human leukaemias and lymphomas using extensive immunophenotypes from an antibody microarray. *British Journal of Haematology*, **135**, 184–197.

26. Zhao, M.P., Li, Y.Z., Guo, Q. *et al.* (2002) A novel scheme for production of polyclonal antibody against estrogenic bisphenols. *Chinese Chemical Letters*, **13**, 845–848.

27. Smith, J.E., Medley, C.D., Tang, Z. *et al.* (2007) Aptamer-conjugated nanoparticles for the collection and detection of multiple cancer cells. *Analytical Chemistry*, **79**, 3075–3082.

28. Tang, Z., Shangguan, D., Wang, K. *et al.* (2007) Selection of aptamers for molecular recognition and characterization of cancer cells. *Analytical Chemistry*, **79**, 4900–4907.

29. Herr, J.K., Smith, J.E., Medley, C.D. *et al.* (2006) Aptamer-conjugated nanoparticles for selective collection and detection of cancer cells. *Analytical Chemistry*, **78**, 2918–2924.

30. Dainiak, M.B., Plieva, F.M., Galaev, I.Y. *et al.* (2005) Cell chromatography: separation of different microbial cells using IMAC supermacroporous monolithic columns. *Biotechnology Progress*, **21**, 644–649.

31. Long, J.L. and Winefordner, J.D. (1983) Limit of detection, a closer look at the IUPAC definition. *Analytical Chemistry*, **55**, 712A–724A.
32. Matsunaga, T., Hosokawa, M., Arakaki, A. *et al.* (2008) High-efficiency signle cell entrapment and fluorescence *in situ* hybridization analysis using a poly(dimethylsiloxane) microfluidic device integrated with a black poly(ethylene terpthalate) micromesh. *Analytical Chemistry*, **80**, 5139–5145.
33. Johnson-White, B., Buquo, L., Zeinali, M., and Ligler, F.S. (2006) Prevention of nonspecific bacterial cell adhesion in immunoassays by use of cranberry juice. *Analytical Chemistry*, **78**, 853–857.
34. Shirazaki, Y., Tanaka, J., Makazu, H. *et al.* (2006) On-chip cell sorting system using laser-induced heating of a thermoreversible gelation polymer to control flow. *Analytical Chemistry*, **78**, 695–701.
35. Fu, A.Y., Chou, H.-P., Spence, C. *et al.* (2002) An integrated microfabricated cell sorter. *Analytical Chemistry*, **74**, 2451–2457.
36. Diez, C. and Simm, A. (1998) Gene expression in rod shaped cardiac myocytes, sorted by flow cytometry. *Cardiovascular Research*, **40**, 530–537.
37. Springston, S.R., Myers, M.N., and Giddings, J.C. (1987) Continuous particle fractionation based on gravitational sediment in split-flow thin cells. *Analytical Chemistry*, **59**, 344–350.
38. Benincasa, M.-A., Moore, L.R., Williams, S. *et al.* (2005) Cell sorting by one gravity SPLITT fractionation. *Analytical Chemistry*, **77**, 5294–5301.
39. Wang, X.-B., Yang, J., Huang, Y. *et al.* (2000) Cell separation by dielectrophoretic field-flow fractionation. *Analytical Chemistry*, **72**, 832–839.
40. Gunasekara, N., Olson, K.J., Musier-Forsyth, K., and Arriaga, E.A. (2004) Capillary electrophoresis separation of nuclei released from single cells. *Analytical Chemistry*, **76**, 655–662.
41. Hu, S., Zhang, L., Krylov, S., and Dovichi, N.J. (2003) Cell cycle-dependent protein fingerprint from a single cancer cell: image cytometry coupled with single-cell capillary sieving electrophoresis. *Analytical Chemistry*, **75**, 3495–3501.
42. Sims, C.E., Meredith, G.D., Krasieva, T.B. *et al.* (1998) Laser-micropipet combination for single-cell analysis. *Analytical Chemistry*, **70**, 4570–4577.
43. Armstrong, D.W. and He, L. (2001) Determination of cell viability in single or mixed samples using capillary electrophoresis laser-induced fluorescence microfluidic systems. *Analytical Chemistry*, **73**, 4551–4557.

6

Flow Cytometry: Cell Analysis in the Fast Lane

6.1 INTRODUCTION

Flow cytometry, like modern microscopy, was developed to fill a technology gap in cell analysis. In fact, the two techniques share a common development and often share similar components, albeit in different geometries. Before one can appreciate the power of flow cytometry, it is beneficial to first look at the inherent limitations of optical microscopy. The microscope (Chapter 4) offers spatial information, a degree of temporal information, and chemical information (if probes and labels are used). However, as the microscope field of view increases or the resolution decreases, the number of cells counted increases. In the extreme cases of high resolution, such as confocal or epifluorescence with $100\times$ objectives, the spatial detail limits the number of cells in a given image. In order to generate a statistically relevant sample, one must take hundreds or thousands of images and analyze them. This task represents a manpower limit on producing statistically accurate data (see Chapter 8 for a discussion of statistical analysis of cells). If lower magnifications are used, then the spatial information content decreases, but cell counts increase. The chemical and temporal information remain largely unchanged, but a loss of morphological information is traded for better statistics. If this progression were taken forward to the limit of no morphological information, then the cell numbers would increase into the thousands, or hundreds of thousands, per image. In later sections of

Practical Cell Analysis Dimitri Pappas
© 2010 John Wiley & Sons, Ltd

this chapter nonflow methods capable of high cell counts using megapixel CCD (charge coupled device) cameras, are discussed. When flow cytometry was originally developed, high-density imaging array detectors like CCD cameras were unavailable, and the best option to measure a vast number of cells – without morphological information – was to flow the cells past a point detector. The laser is an ideal light source in this case, since it can be focused to a small area and is intense and monochromatic. This essential combination of flow system, laser(s), and optical detectors form the basis of flow cytometry. Virtually any fluorescence assay conducted on a microscope can be analyzed by flow cytometry – with the exception of spatial/morphological measurements – with cell counts that approach the millions in some cases. This chapter is meant to serve as an overview of flow cytometry and a practical set of tips to start working with cytometers for cell analysis. For a fascinating historical and technical account of flow cytometry development and usage, the reader is directed to Howard Shapiro's *Practical Flow Cytometry* and several other texts [1–4].

Flow cytometry is ideally suited to cell analysis since it was developed for that purpose and provides high cell counts. When measuring any population, whether it is comprised of cells or molecules, and so on, it is always preferable to count individual responses and assemble a histogram or table comprised of individual values rather than to measure cells as an ensemble. For example, microplate assays for cell viability (using a fluorescent DNA stain) will yield an average fluorescence value for the cell population. However, if the fluorescence intensity for the average is 30% higher than the controls, what conclusion can be drawn about the sample? The fluorescence is, of course, larger, so some amount of DNA is being stained. Are 30% of the cells in the sample dead? Are a majority of the cells partially stained (e.g., apoptotic)? By taking an ensemble reading, the true state of the population can be obscured. Flow cytometry, however, can measure tens of thousands of cells in a short analysis time, and the viability of each cell is recorded. Making stochastic measurements in this manner results in an accurate representation of the sample. The mean fluorescence of the population may be 30% brighter than the controls, but one can now identify *how many* cells are dead, or apoptotic, or viable. This sample resolution beyond ensemble averaging is one of the true benefits of flow cytometry, combined with large cell counts for statistically representative samples.

Flow cytometry is also an attractive technique because multiple parameters can be measured simultaneously for each cell. In epifluorescence microscopy, in most cases each fluorescent probe or label requires an additional exposure to acquire each wavelength. This limitation is not present in most laser-scanning confocal instruments, but even these high-end instruments detect 3–5 parameters in most cases. The simplest flow cytometers on the market offer 3–4 fluorescence channels, plus forward and side scatter, resulting in 5–6 parameters per cell. Herzenberg and colleagues at Stanford and the National Institutes of Health continue to push the barriers of multi-parameter measurements, exceeding 13 fluor-escence colors and two scatter measurements with customized, multi-laser systems [5]. Several commercial systems now offer 18 fluorescence colors or more. Of course, expanding beyond 4–5 fluorescence parameters complicates instrument setup, the number of controls, and compensation, but is sometimes necessary and oftentimes desired. In one measurement, one can assess viability, phosphotidyl serine exposure, and several fluoro-genic caspase probes in apoptotic cell samples with the most basic cytometers. On the high end, complete multi-immunophenotyping of lymphocyte sub-sets is possible [6]. Flow cytometers can also be equipped with sorting mechanisms for cell separations (Chapter 5). It is this information-rich approach, combined with live cell measurement cap-abilities, that make flow cytometry one of the most powerful cytometric approaches available.

For all of the benefits of flow cytometry, there are some limitations to the technique. With the exception of the Amnis Imagestream instrument, commercial flow cytometers do not provide morphological information, which is necessary for certain analyses. Also, flow cytometers are ex-pensive compared to most microscopes. The upkeep of the instrument often precludes each investigator from having his or her flow cytometer for personal use. Most importantly, flow cytometers are prone to many subjective and systematic errors on the user end. There are many tricks to getting good flow data, and many opportunities to ruin a good experiment with bad flow cytometry protocols or approaches (Table 6.1). The aim of this chapter is to bring out some of these subtle pitfalls and allow researchers new to flow cytometry to obtain data that is as good as the cell sample allows.

6.2   THE CELL SAMPLE

In flow cytometry the sample is the limiting portion of the experiment. A poorly characterized or understood sample will never yield satisfactory results. When core facilities are used, it is especially important to under-stand the sample well, so that shared instrumentation is not damaged, or

Table 6.1   Troubleshooting flow cytometry

No cells are present (before adjusting thresholds/compensation)	Threshold may be too high, or the flow chamber is clogged. Lower threshold until events appear (if no events, then a clog is suspect). Check with beads to assure flow is occurring
No cells are present (correct thresholds/compensation)	A clog has occurred, or the cell concentration is too low. Run a sample of test beads to ensure the cytometer is working properly. If the cytometer is functional, check the sample on a microscope
Cells "Disappear" from dot plot after setting detector gain.	Cells are "compressed" in the lowest or highest channel (value) of the data plot. Adjust detector sensitivity.
Cells "disappear" after compensation	See above. Lower compensation settings
Too many cells detected (double occupancies)	If this is determined during measurement, dilute the cell sample. If discovered after the experiment, use scatter channels to gate only single occupancies
Single-color controls appear in two fluorescence channels	Use compensation to reduce fluorescence bleed-through
Signal from unstained/isotype controls too high	Wash cells to reduce background fluorescence, or adjust detector gain
Sudden appearance of thousands of events with random scatter/fluorescence values	Bubbles have been drawn into the system. Replace sample, flush system and repeat analysis

at least clogged to the point that no one else can use the cytometer for the day. Sources of sample error are many and varied. The first step to measuring cells on a flow cytometer begins on the microscope. It seems counterintuitive, but a quick glance at the cells under the microscope will determine if there are clumps of cells, poorly stained cells, dead cells, or any cells at all. Counting the cells on a hemacytometer (Chapter 2) is also recommended, as most flow cytometers do not determine cell concentrations.

Cell composition and quality are the most important aspects to good flow data. If the cells are fixed and stained, then errors in the fixation or staining steps will affect the overall experimental outcome. If the cells are viable, there are more obstacles to consider. Getting the cells to the cytometer in a timely fashion is vital to maintaining viability and not altering the cells. Of course, if one owns the cytometer, this trip is usually not an issue. However, if the flow cytometer is 20 minutes drive through traffic, an incubated cell container is recommended. A heating pad in a Styrofoam cooler usually suffices in this case. Transporting the cells on ice is also possible. In either approach, care must be taken to maintain viability and function, otherwise the experiment will fail.

A seemingly obvious caveat, yet one that is overlooked too often, is that the cells should be small enough to exit the cytometer flow chamber and orifice (if it is a sorter) without clogging. Cells should not be clumped together, and the sample concentration should be high enough to allow larger cell counts (shorter analysis), but not jam the flow chamber. In the latter case, many flow users believe concentrated is better; this is not true. An overly concentrated sample leads to double occupancies (Section 6.5) and may clog the cytometer. If the cells are too large, a different cytometer or sorter must be found. Clogging is not a fatal error, but one that may earn one the ire of the flow cytometer owner/keeper. Clogging will, at the very least, waste a sample and slow down the analysis. Often, flow experiments are time consuming to prepare, and many samples or time-scales means long analysis times. To develop a clog in the middle of such an experiment results in an entire day of research wasted. Some clogs can be removed through the cytometer fluidic controls. For example, the FACSCalibur instrument from Becton-Dickinson has a purge control that usually removes any small clogs. However, sometimes the clog is severe enough to require extensive rinses with dilute bleach or pure water (it is best to check the documentation of the instrument for this level of cleaning). Another sample issue that affects the outcome of flow experiments is the staining of the sample itself. With the exception of cell conductivity, most cell measurements in flow cytometry require fluorescence staining of some sort. Just as in microscopy, poor staining procedures can doom what was otherwise a good sample. Cross talk between fluorophores is more problematic in flow cytometry than in microscopy. Scientists rarely attempt to measure more than three or four parameters by microscopy, yet these same individuals routinely utilize every last wavelength to measure as many parameters by flow cytometry as possible. Proper staining techniques for flow cytometry are discussed in Section 6.6.

6.3   FLOW CYTOMETER FUNCTION

Figure 6.1 depicts a typical flow system and detection apparatus. All flow cytometers contain a fluid system, a flow cell, and detection optics (or electrodes, in the case of impedance measurements). Sorting mechanisms are discussed in Chapter 5. For the sake of simplicity, the exact details of each instrument on the market today are not discussed; rather, a general discussion of flow cytometer function is given. The user should be able to

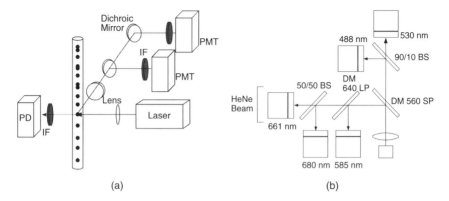

(a)   (b)

Figure 6.1   (a) Conceptual diagram of a fluorescence-based flow cytometer. Laser excitation is used to measure forward and side scatter, as well as several fluorescence emission wavelengths simultaneously (IF = interference filter; PD = photodiode; PMT = photomultiplier tube). (b) Optical layout of the FACSCalibur 4-Color Flow Cytometer. Fluorescence bands are separated by dichroic mirrors (DM = dichroic mirror; SP = short pass, LP = long pass). Interference filters define wavelength bands. A second laser is used to excite an additional detector in the 660 nm region of the emission spectrum. More complex flow cytometers use a similar system for wavelength selection and detection

select which instrument to use or buy after reading this chapter. Any cytometer begins with the fluid system, a mechanism to drive fluid flow through the measurement region. In most commercial flow cytometers, compressed air is used to drive fluid flow. The compressed air method is efficient and inexpensive, and allows for high, sustained flow rates. Both the sheath and sample streams can be driven by this method. The downside to compressed air flow is that the flow rates are not uniform, and no volumetric measurements are possible. In most cases, the volumetric flow rate (i.e., volume consumed) of the sheath fluid is not important. The sample volume consumed, however, can be used to calculate the concentration of the detected cells (a useful parameter normally determined by hemacytometer). A simple method to calibrate sample flow rates is to fill a sample tube with water and run the system at the desired speed for as long as possible. Weighing the tube before and after the run will determine the weight (and hence volume) consumed over the run time. The longest possible run time ensures the most accurate volumetric flow rate. This calibration should be performed at each sample flow rate, and should be repeated whenever the instrument's flow system is adjusted, or at least semi-annually.

Gravity flow can be used in the slowest flow systems [7]. Gravity flow is advantageous for field measurements, since it requires no power, and can be adjusted by altering the height of the sheath reservoir relative to the flow chamber. Gravity flow results in slower flow rates. For faster flow, peristaltic or syringe pumps can also be used as alternatives to pressurized flow. Peristaltic pumps are advantageous in that they have no volume limit (i.e., they can run until the reservoir is empty). Peristaltic pumps, however, suffer from pulsatile flow and cannot match the flow rates of pressure-driven flow. Using pulse dampeners or two peristaltic pumps out of phase with each other can reduce flow pulsing, but the required complexity of these approaches negates any advantage over pressure-driven flow. Syringe pumps, however, have many advantages over other flow methods. They are truly volumetric, so that accurate cell concentrations are possible – provided that the cell concentration is not too high. Syringe pumps also offer smooth, nonpulsatile flow and in some cases fast flow rates. The main disadvantage of syringe pumps is that the delivered fluid volume is limited by the syringe size. Some commercial systems use a syringe pump to deliver sample and compressed air to drive the sheath fluid. This approach yields volumetric metering of the sample combined with high flow rates and simplicity for the sheath flow.

Most flow cytometers are driven by air pressure, and the flow from the sample into the flow chamber is not quantitative. Simple methods to calibrate pressure-driven flow are discussed above. However, even if the flow is calibrated, variations in flow rate in pressure-driven systems mean that an external method of calibration is needed. There are several commercial counting beads available. The premise behind these calibration beads is that they are of known concentration, so a known aliquot can be used to measure the sample volume. The concentration calibration beads are typically of much higher intensity, so that they can be gated independently from the cells in the sample. Even with syringe systems, it is best to calibrate the system using counting beads periodically.

The flow chamber itself is in many ways the heart of the flow cytometer. It is at this point (see Figure 6.2) that cells are aligned in the flow streams and measured. The purpose of the flow cell for most instruments is to align all cells so that they share a common speed and trajectory through the measurement volume. For example, if a laser is used to interrogate cells as they pass through the flow chamber, then cells that interact with the edges of the beam (i.e., not in the center of the stream) may experience reduced excitation efficiency and lower signal. A cell traveling at lower speed than its neighbors may produce a larger signal than cells traveling faster. The result of these variations in flow and position is a cell population whose fluorescence values are broadened by, not only biological and chemical statistics, but also poor instrument precision. The use of sheath fluid

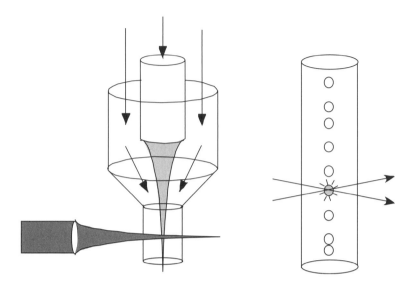

Figure 6.2 Conventional flow system in a flow cytometer. A concentric flow of sample (inner stream) and sheath fluid reach a neck-down region where the flow chamber diameter decreases before the laser beam excites the cells (Left). The resultant flow is laminar, and cells are focused to a single-file stream with nearly identical velocities

therefore normalizes cell passage through the flow cell and is critical for precise measurements. While many flow chamber geometries exist, they can all be grouped into sheathed or sheathless systems. In the former, concentric sheath and sample streams compress spatially (Figure 6.2) to produce laminar flow and a hydrodynamically focused sample core. Sheathless systems – the most notable is available from Guava Technologies – do not use sheath fluid, and drive flow using a syringe pump. Sheathless systems use less reagent volume per analysis and are easy to use and maintain, but the aforementioned loss in precision may introduce unnecessary error in the application. One must evaluate if the precision of sheathless systems are sufficient for the task at hand. In many cases, such as DNA content measurement, the higher precision offered by systems using sheath flow are necessary.

The flow chamber itself accomplishes many tasks simultaneously. First, it focuses cells hydrodynamically (if a sheath is used) and is the interface to the detection mechanism. If impedance measurements are made, a Coulter orifice is present in the flow chamber to measure cell size and count cells. For fluorescence measurements, the laser source(s) can pass through the flow chamber itself or into the stream after the flow nozzle (if the instrument is a stream-in-air sorter). The flow chamber must be made of high quality optical materials for fluorescence measurements to mini-mize scatter. The Guava flow cytometer is an exception, since the flow chamber is a square glass capillary with the polyimide coating removed for optical transparency. If the flow chamber clogs, the user simply replaces it in the Guava flow cytometer. Severe damage to the flow chambers of other flow cytometers results in costly repair, which is one case where the Guava systems have a clear advantage.

Cell detection by impedance is one of the oldest approaches to cell measurements in a cytometer. There are two main types of electrical measurement, DC resistance and AC impedance. The difference between the two is the frequency component of the AC field. It is also possible to measure the electrical opacity of a cell, which is the ratio of AC impedance to DC resistance. DC and AC measurements are used to both count cells and determine their size. Opacity can be used to identify cells of similar volume, but different dielectric characteristics. The fundamental concept of the Coulter orifice is that the cell is a poor conductor of electricity relative to saline. As a cell passes through the orifice (Figure 6.3) the displacement of saline increases the resistance of the orifice. The relationship between the cell size and resistance change is given by the

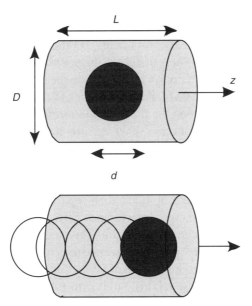

Figure 6.3 Passage of a cell through a Colter orifice. The cell diameter (and hence the volume) affects the displacement of conductive buffer, resulting in an increase in orifice resistance. The effect is transient (bottom), as the cell cross-sectional area changes as it traverses the orifice length (L) along the fluid axis (z)

following [8],

$$V_P = \frac{A^2}{r_0} \Delta R \, f(\alpha)$$

where α is the ratio of particle cross-section (maximum) to orifice cross-section, V_p is the particle volume, r_0 is the buffer resistivity, and ΔR the change in resistance (pulse height), and $f(\alpha)$ is a correction factor between 0 and 1.0. The orifice cross section is given by A.

The Coulter orifice is effective if cell counting and a degree of sizing information are desirable. There are, however, some limitations to the technique. Measuring bacteria in conventional Coulter counters is a challenge, since the cross-sectional area of the bacterium is small compared to the orifice. Also, cells of similar volume are difficult to distinguish, and cells with large aspect ratios (such as erythrocytes) can yield different values, depending on orientation throughout the orifice. Nevertheless, DC and AC measurements of cells are rapid, inexpensive relative to fluorescence, and are often preferable for cell size measurements.

Fluorescence measurements are needed for more robust measurements of cell type or to monitor cellular mechanisms. The fluorescence detection system shown in Figure 6.1 is a simplified, multi-parameter system that is found in most instruments. The number of fluorescence detectors varies by instrument, but the essential components are the same across platforms. The laser is focused into a sheet of light using prisms or cylindrical lenses to ensure that cells passing through it experience the same irradiance. When multiple lasers are used, as is common in higher-end instruments, the lasers are separated spatially along the flow axis. Fluorescence signals from cells as each passes through successive lasers are delayed in time, but are combined so that the fluorescence signal from each laser excitation is assigned to the correct cell. Fluorescence from the cells is collected by an objective or high numerical aperture lens and transmitted to the detection system. Bands of fluorescent light are then separated by dichroic mirrors and directed to each detector. Interference filters ensure minimal interference between fluorescence bands, but this setup limits experimental flexibility to a degree. Like most fluorescence microscopes, the use of fixed-wavelength filters allows high transmission, but restricts the available fluorophores that can be used. Fortunately, most flow cytometers have – at the least – an argon-ion or solid-state laser operating at 488 nm, and most companies selling flow cytometry reagents match their fluorophore-conjugated reagents to match most instruments. However, if a one has to expand beyond the typical palette of reagents, the fluorophore must be matched to the detectors. Detection in flow cytometry is usually performed with photomultipliers, although deep-red fluorophores are also detected by avalanche photodiodes. The gain of the multipliers is controllable in most cases so that signals can be adjusted for the required sensitivity.

Flow cytometry produces a population of individually measured cells. There are multiple ways to display and analyze data, depending on the experiment. If only one parameter is measured, a histogram of fluorescence intensities (of each cell) is the simplest method of display and analysis (Figure 6.4). Gates drawn around the histogram peaks generate

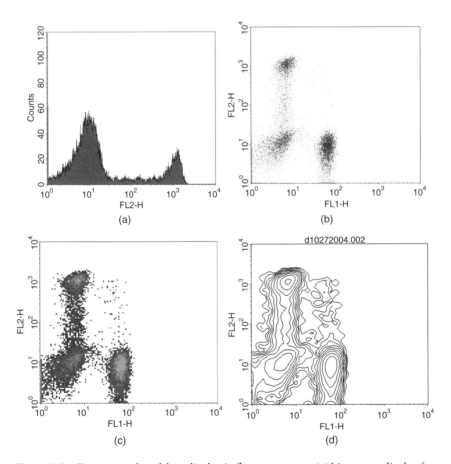

Figure 6.4   Four examples of data display in flow cytometry: (a) histogram display for single parameter measurements; (b) dot plot (each dot represents one cell), suitable for display on screen, but lacks density information; (c) pseudocolor density displays, suitable for presentations or publications, but carry little meaning where color is not allowed; (d) contour plot (5% contours shown); suitable for black and white publication

statistics about each population (see Section 6.7). If more than one parameter is measured for each cell, then a 2D plot is preferred. There are several types of 2D plots (Figure 6.4), the dot (or scatter) plot being the simplest. Each point on the plot represents a cell, and this display can be thought of as a 2D histogram. When acquiring flow-cytometry data, one can often choose how many dots are displayed. For example, if 2000 dots are displayed, then the 2001st dot replaces the first, and so forth. All of the cells are acquired and are displayed during the analysis, but this truncated data approach is useful for setting up parameters (Section 6.6), as changes in the signal are readily observed on screen. Dot plots are amenable to publication as they are monochrome. The key drawback of dot plots is that once the dots in a particular area saturate the image, then no additional quantitative observations can be made. Of course, analyzing the file using gates will yield the correct cell number, but *visual* observation is limited when saturation occurs. To avoid saturation, a density plot can be used. Density plots are similar to dot plots, but a color gradient is added to indicate increasing density of dots. Saturation is eliminated in this case, but each cell is still represented by a dot, allowing for outlier cells to be observed. Density plots are typically grayscale. Pseudocolor plots are colorized density plots. Pseudocolor plots are vivid for presentations, but are otherwise identical to density plots. Contour plots are 2D plots with each contour representing an incremental increase in cell number. Contour plots are preferable for many publications, as they accurately represent the cell sample in a simple format. If outliers can be plotted along with the contours, then rare cells can also be depicted in this plot. Contours are set up as increments in the percentage of the total cell population (typically 5%). The number of contour increments chosen will affect how clean data appears, but this number should be adjusted by need (i.e., publication vs. presentation).

It is also possible to smooth data obtained by flow cytometry. Histograms and 2D plots can all be smoothed by some amount (typically 1–5%). In general, data smoothing can remove some of the noise introduced by the instrument or the sample. However, it is generally not a good idea to smooth data arbitrarily. Smoothing that occurs by averaging multiple measurements is valid, but choosing a smoothing parameter to make data look "cleaner" is misleading. It is always preferable in any analysis to present the results as-is, to let the experiments speak for themselves. Since smoothing is subjective at best and misleading at worst; it should be avoided unless it is essential to the data analysis process.

6.4   OBTAINING OR FINDING A FLOW CYTOMETER

Flow cytometers are not instruments generally found in every laboratory. The cost of these instruments (particularly sorters) must be weighed against the data one will obtain from it. Even the so-called low-cost instruments can cost upwards of US$70 000, which is not a small sum of money considering today's funding climate. A new investigator with set-up funds may want to consider buying their own flow cytometer, if it will be used daily and will generate results that will in turn generate grant funds. If cytometry use will be less than 3–4 times each week, it is better to seek out a core instrument or collaborate with someone who has a flow cytometer. Flow cytometers are often found in biology departments or medical colleges. Most hospitals have several flow instruments, but are unlikely to allow outside individuals access. One of the best ways to find a flow cytometer nearby is to join Purdue University's flow cytometry web page. The discussion boards are populated with knowledgeable and helpful people, and a nearby flow expert may be a Web search away.

Before purchasing or finding a flow cytometer to use, one must consider the types of instruments to find the correct one. The first question that must be addressed is if sorting is needed for a given experiment. If so, the number of instruments and their availability decrease. Flow instruments are generally fluorescence based and capable of high-speed cell analysis and sorting. Most are stream-in-air sorters, although the Partec instruments and Becton-Dickinson's FACSCalibur use fluidic sorters. The benefit of using a sorter is that most have more than five fluorescence parameters. If sorting is not needed, then any flow sorter or analyzer instrument can be used, and the probability of finding an available instrument increases.

The next decision to make is whether the measurement method will be electrical or fluorescence based. The outcome of this decision should be immediately clear. If cell counting is needed, either method will work. If cell size information is needed, electrical measurements are preferred. Fluorescence measurements are, of course, more flexible and will most likely be the method of choice for most users. The fluorophore to be used with the cell sample must match one of the instrument's excitation lasers and one of the photodetectors. For example, if the flow cytometer uses a single-line argon-ion or solid-state laser at 488 nm, then any dye used must absorb at that wavelength. Most of those dyes will therefore emit near 520–540 nm, limiting the number of detector channels. For the venerable 488 nm line of the argon-ion (and its newer, solid-state replacements), many dyes have been developed so that more than one detector channel can be used. For example, it is possible to use a fluorescein or rhodamine derivative for the first fluorescence channel (FL-1 in most instruments), phycoerythrin for FL-2, and propidium iodide or 7-AAD for DNA measurements (FL-3 in this case). Using tandem conjugates of phycoerythrin or other fluorescent proteins extends the excitation capabilities of the 488 nm line into the far-red region of the electromagnetic spectrum. It is therefore possible to measure four or five fluorescence parameters with a single laser line. Most antibodies are available with fluorophore conjugates that match flow cytometer detector channels. Custom probes or fluorescent proteins must be chosen with care to ensure proper performance in the cytometer.

If purchasing a flow cytometer, one must forecast what possible analyses will be performed during the lifespan of the instrument. Only those flush with set-up funds or recipients of very large grants can afford instruments for their own laboratories. Core facilities are more likely to receive funding for instruments of this cost. In either case, a good plan of what will be performed and how often, not only helps the cytometer user, but also makes a compelling case when seeking funds for a cytometer. For example, if the instrument is for a particular lab, how many fluorescence probes will be assayed simultaneously? Not every investigator needs to measure 12 antibodies at the same time. Some instruments, such as the Influx sorters and analyzers are customizable. Other instruments have less flexibility, although it is possible to outfit many systems with an additional fluorescence channel. The simplest instruments typically feature three fluorescence bands and one or two scatter channels (forward and, at times, side scatter). A four- or five-color instrument is likely to meet current needs for most users. It is inadvisable to buy the most expensive instrument just because the funds are available. A nine-color instrument with sorting functions is a fine instrument, unless the owner finds that most analyses only take advantage of three detector channels and that the lab only performs four sorting experiments a year. By overestimating what capabilities will be needed, but not choosing unnecessary features, the lab's funds can be used most efficiently.

6.5   USING FLOW CYTOMETERS

Sample handling is the first and greatest source of systematic error in any flow-cytometry experiment. Cell samples can vary by species, by source (i.e., cultured vs. primary), and by concentration. Prior to any experiment, one must consider the safety aspects of flow cytometers. The lasers and electronics are usually shielded from the user, so the main hazard is biological (Chapter 2). If the instrument is in a shared facility, or if the investigator generously allows others to use his or her instrument, then it is up to the owner of the instrument to ensure that all specimens brought into the lab comply with proper safety. Training of users in blood-borne pathogens is recommended, and often required.

The sample composition can vary as well. If the cells are cultured, then there is little variation except for cell size, possible clumping, and so on. Many cancerous lymphocytes available for culture exhibit clumping behavior, and must be filtered or the aggregates must be disrupted prior to analysis. Primary cells, however, present different challenges. Blood is one of the most common cytometry samples, and presents unique challenges. As discussed in Chapter 5, lysis of blood is often desired, as the large number of erythrocytes can obscure other cells. An alternative approach to blood measurement is to use "immunogating" where anti-CD45 fluorescence and side scatter are used to analyze only leukocytes, omitting the lysis step altogether. Blood also has a high protein content and is generally presents more clogging problems than cultured cells. Extracellular matrix from isolated primary cells (prior to sorting) can also cause problems with the fluidics, as well as the scatter and fluorescence signals. Primary cells are also more problematic, since the initial cell harvest may contain unwanted cell types and bacteria.

One of the hidden pitfalls during flow measurements is disregard for the sample concentration. At the extreme, low concentrations result in long analysis or sort times, and also make setup difficult. When adjusting the photomultiplier tube and forward scatter detectors, and also when setting up compensation, a high cell concentration ensures immediate update of the data displays. This fast feedback allows settings to be adjusted quickly. If the cell concentration is low, then one must wait until enough cells are counted with the new instrument setting to observe the results. Flow cytometry is a visually oriented analysis technique, as the user establishes the settings and gates after visual inspection of the data. If multiple parameters are measured, then each must be set up by inspecting the data and adjusting detector gain, and so on. For example, if immunogating is used to count CD8 + lymphocytes in blood, then at least three fluorescence parameters (detectors for anti-CD45, anti-CD3, and anti-CD8) and two scatter channels may be measured simultaneously. These five parameters must individually be optimized using a control sample before actual analysis occurs. In addition, compensation between the three fluorescence channels may be required, resulting in a total of seven or

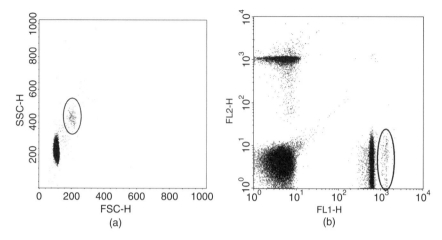

Figure 6.5 Double occupancies due to high cell concentration: (a) beads showing double occupancies (in the oval gate). Double occupancies in this case have roughly twice the side scatter and forward scatter intensities; (b) effect of double occupancies on fluorescence signals, shown with beads. The oval gate shows the double occupancies, which appear to have twice the fluorescence. Gating only the single occupancies removes this source of error

eight parameters to adjust before the analysis can begin. If the cell count is too low, the total setup time will be lengthy.

At high cell concentrations clogging or double occupancies may occur. Double occupancy occurs when the concentration is high enough to ensure that two cells arrive at the same time in the cytometer (Figure 6.5). At any point in time there will be n number of cells in the laser beam ($n = 0$, 1, 2...). At low concentrations, most of the time no cells will be present. When the concentration increases, the probability that two cells will be in the probe volume increases. The scatter and fluorescence signals will be larger for these so-called double occupancies, and could potentially be eliminated by proper gating. However, if one is unaware that double occupancies are occurring, many artifacts may be introduced into the data set. For example, if two cell types (one expressing antigen A, one expressing antigen B) are mixed and measured at high concentration, then three populations will result. There will be two large populations (cell A and cell B, respectively) plus a smaller cell population that appears to express both antigens A and B. The true cause of this third, erroneous population is the double occupancy of different cell types.

Tip: It is critical that cell concentration be adjusted so that the analysis time is acceptable, but clogs and multiple occupancies are avoided.

In the latter case, a quick dilution with saline will reduce the concentration to an appropriate level. In case of the former, centrifugation and resuspension will be needed to increase cell concentration.

6.6   SETTING UP A FLOW CYTOMETER FOR MULTI-COLOR STAINING

Whether one or more parameters will be measured by flow cytometry, careful adjustment of the instrument settings is required for successful results. For each parameter – whether scatter or fluorescence – a detector voltage or gain must be set. In addition, compensation between detector channels may also be required, resulting in additional adjustments. Finally, thresholds may also be set up so that cells falling below a certain measured value (e.g., scatter) are rejected. If gating is required for data acquisition or sorting, then additional parameters must be set. This multi-parameter adjustment occurs for each set of experiments, and requires proper controls. To illustrate this concept, it is best to use a test case. For example, if lysed blood is to be analyzed for CD3 + CD4/CD3 + CD8 + T cell ratios, then several different protocols can be used. First, anti-CD45 (for this case attached to fluorescein) and forward and side scatter may be used to identify leukocytes in the lysed sample. Second, an anti-CD3 stain (e.g., using allophycocyanin, APC, as the fluorophore) will mark all T cells. Antibodies for CD4 (conjugated to phycoerythrin) and CD8 (conjugated to peridinin chlorophyll A protein (PerCP)) could then be used to count the anti-CD4 and anti-CD8 T cells, respectively. To perform this analysis, the flow-cytometry instrument would require two lasers (an argon-ion at 488 nm and a diode laser at 635 nm), four fluorescence detectors, and two scatter channels. Each channel (six in total) would require voltage or gain settings and compensation between the channels. To properly set up parameters, the following controls would be needed.

1. Unstained sample (as a check for autofluorescence)
2. A control stained only with anti-CD45 FITC (fluorescein isothiocyanate; fluorescein)
3. A control stained only with anti-CD3 APC (allophycocyanin)
4. A control stained only with anti-CD4 PE (phycoerythrin)
5. A control stained only with anti-CD8 PerCP
6. Samples stained with all fluorophores (the analytical sample).

Samples 1–5 would be needed to set up the instrument and also ensure that proper compensation takes place. In the case where off-line compensation is performed in software, then these samples would also be measured to set up the compensation matrix. Sample 6 is the analytical sample to be measured, and, in practice, this sample set comprises all of the analyses to be performed in that experiment. Many of the settings could be used repeatedly for weeks or months, but it is best to calibrate the instrument with each use. It is easy to see that deviation from this simple example into more complex staining requires more controls and careful experimental design.

There are three main groups of parameters to set in flow cytometry: the threshold(s), the detector sensitivities, and any compensation. Typically, the threshold and sensitivity of the trigger channel – typically forward scatter – are set together. For most cases, an unstained control is ideal for setting thresholds. Thresholds are used, particularly with forward scatter, to remove any particles that are too small to be cells (i.e., debris). This is important, because, while flow cytometers run continuously during sample analysis, data is only collected when a trigger (usually forward scatter) indicates a cell is present in the measurement volume. Counting debris or smaller particles therefore results in most of the detected population consisting of useless events. This is especially critical when an experiment is set up to count a specific number of particles (events) and then stop. For example, if measuring cultured cells, debris and remnants of cells may comprise a significant portion of the sample. Using a threshold to eliminate these unwanted events allows the instrument to only record cells of interest. Setting the threshold too low (Figure 6.6) results in many events detected per second, but the majority of these events are meaningless. Too high a threshold (Figure 6.6) results in many target cells going undetected, and leads to excessive analysis times. It is possible to also gate using more than one threshold parameter, such as forward scatter (for size) and intensity of a DNA stain (to accept only nucleated cells). This example is used in Guava's viability assays, where cell debris is rejected using a permeant DNA stain and scatter to reject non-nucleated particles. A second, impermeant DNA stain then identifies dead and live cells based on fluorescence intensity. In the Guava software, the permeant DNA stain threshold ignores events below the threshold value (chosen by the user) so that the resultant data file only contains cells of interest.

For scatter measurements, either linear or log-scale measurements can be made, depending on the analysis. Linear measurements of forward scatter are better for viewing smaller changes in particle size (and may be used to identify double occupancies, which will be approximately twice as large as single particle events). Many side-scatter channels use

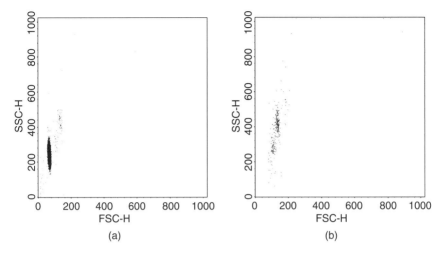

Figure 6.6   Effect of incorrect threshold on flow cytometry analysis: (a) correct threshold, set just below the forward scatter signal of a sample of beads; (b) incorrect threshold, set at 60% of the mean forward scatter signal for the same beads. A high threshold will yield few cell counts, while an excessively low threshold will also detect debris in the sample

photomultiplier tubes, so the voltage to that detector can be adjusted (in software) to control the sensitivity. Forward-scatter channels typically use photodiodes, and have an amplifier with several pre-set gain values to adjust the sensitivity by a factor of two, four, and so on. Once thresholds are established on scatter channels, the sensitivity of the fluorescence channels must be set. Before setting the sensitivity using an unstained control, the choice of log or linear scales must be made. In general, a linear scale should be used for DNA content measurements or any time the change in fluorescence will be small (e.g., doubled, tripled, etc.). Otherwise, a default choice of log scale ensures a larger dynamic range of fluorescence measurements. With the scales chosen, one approach to setting thresholds is to use the unstained control and adjust the fluorescence photomultiplier gain for each channel, so that the unstained control population resides in the first decade of the log scale (i.e., values between 10^0 and 10^1, Figure 6.7). Isotype controls can also be used in place of the unstained control (Section 6.7). Setting the unstained or isotype control in the smallest decade will ensure that the largest dynamic range is possible. With the threshold and sensitivities set, any compensation can be established. Compensation accounts for the overlap between two fluorophores on two detector channels. For example, phycoerythrin will mostly be detected in the FL-2 channel of many instruments, while a portion of the

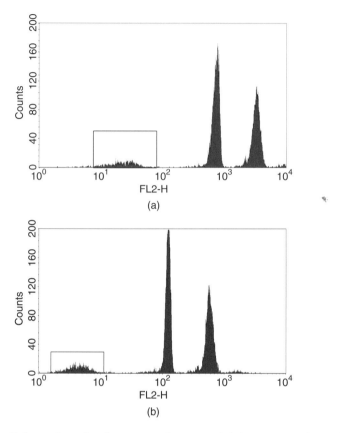

Figure 6.7 Gain settings for fluorescent detectors: (a) incorrect setting, where unstained cells have fluorescence signals between 10 and 100 (second decade); (b) correct setting, where detector gain was lowered to place unstained controls in the first decade (values of 1–10)

fluorescence will be detected in FL-3. Likewise, a smaller portion of that same phycoerythrin fluorescence will be detected by FL-1. Using a sample stained only with the phycoerythrin-conjugated antibody, compensation for Channels 1 and 3 are set. A percentage value of FL-2 is subtracted from FL-1, and a similar subtraction is performed for FL-3. This process is repeated for each label or probe, with the neighboring channels compensated using a singly stained control. The result of compensation is shown in Figure 6.8. Before compensation, fluorescence overlap between channels results in one label or probe giving signals in both channels. After compensation, the signal from one channel only yields signals in the first decade of the other, resembling the unstained control.

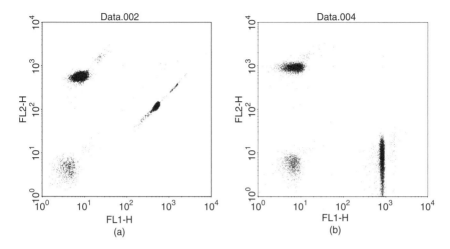

Figure 6.8   Effect of compensation on flow cytometry data; (a) uncompensated data, showing bleed-through of green fluorescence (FL-1) into the yellow fluorescence (FL-2) channel. Bleed through from FL-2 to FL-1 is minimal; (b) compensated data, where the effect of FL-1 on FL-2 signal is subtracted

6.7   ANALYZING FLOW CYTOMETRY DATA

Setting up flow cytometers can, at times, seem as much of an art as it is a science. Once the experiment is finished, and data analysis ensues, there are many opportunities to introduce bias. While the detector voltages, gain, thresholds, and compensation settings can all be set incorrectly and introduce error, the analysis of even correctly set up samples can ruin an experiment. If one considers the many parameters that must be set for analysis, the need for careful – and documented – data analysis procedures is clear. The raw data must often be gated based on one or more parameters. Several other parameters may also be gated, with statistics generated for the gated cells. Even without ever-present subjectivity, slight variation of setting the gates to count cells will introduce errors. Figure 6.9 shows a histogram of cultured T lymphocytes, stained with anti-CD71-fluorescein. This is a sample containing only one cell type, with a single antibody stain for simplicity. The variation in setting the gates will introduce another source of error in the analysis. The error is, in most cases, smaller than the variation of the cell sample itself. However, given that many gates are often set for a single sample, the total amount of systematic error introduced can be large.

    There are several methods available for setting gates of cell populations. The gates themselves serve two roles. First, they count the cells in the region

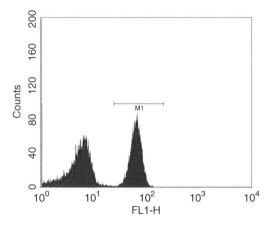

Figure 6.9   Histogram with marker for cell counting. Subjectivity of gate and marker settings is one of the main sources of systematic error in flow cytometry

and assemble statistics on cell number, fluorescence (or scatter) intensity, and so on. The gates can also be used to select a portion of the cells to further analyze cells only in the gate. Doing so separates the gated cells from the "parent" population; for example, separating only proliferating cells for cell-cycle analysis. Cells gated in this manner can be analyzed with respect to only the gated population, or with respect to the entire parent population. For histogram analysis (Figure 6.9), the gate consists of a high and low value (indicated by brackets), where cells out of bounds are not counted/gated. For 2D plots (regardless of plot type), several gating methods exist (Figure 6.10). If the data is well separated – that is, the fluorescence intensities differ between controls and samples – and the fluorescence signals are compensated, then a quadrant can be used to analyze data. Of course, if fluorescence is not compensated, then quadrant analysis is difficult. The benefit of quadrant analysis is that the markers on the x- and y-axes define the minimum signal that is different from a control. Therefore any value to the right of (or above) the quadrant lines are identified as positive for a given fluorescence stain. Since there is no upper bound on either of the two measured parameters, the quadrant analysis is insensitive to differences in intensity for positively stained samples. The other gate types, primarily ellipse and polygon, set both upper and lower bounds on a population. These gates are useful when compensation is not used (or is impossible), or when the differences in cell type are difficult to discern. For example, quadrant gating of blood cells is difficult by scatter, as both upper and lower bounds are needed to isolate lymphocytes and monocytes. Ellipse gates are simple to set up, but are limited when the

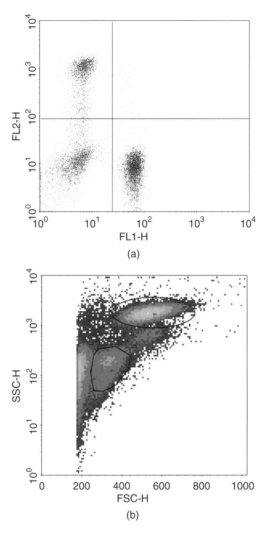

Figure 6.10   Gating methods in flow cytometry: (a) quadrant analysis is useful for well-compensated 2D data; (b) for uncompensated data, or data that is not amenable to quadrant analysis, polygon or oval-shaped gates can be drawn for analysis or gating

population to be gated is irregularly shaped, or when the cell populations are adjacent (or overlap). Polygonal gates allow for greater flexibility, and allow for multiple gates to be fitted around adjacent cell types.

Setting gates is best accomplished using either unstained or isotype controls. The latter is a time-honored, yet still controversial, approach. The isotype control is an antibody of the same antibody type, but that doesn't recognize the antigen of interest. The isotype control should have the same fluorophore as the analytical antibody to correct for the intensity of a cell that does not bind to the analytical antibody. While this method is often cited as the best method of control, there are some concerns with isotope controls. First, if the isotype antibody binds to some other molecule on the cell, is the isotype indeed a control? If this is the case, then what does the isotype correct for? If the cell sample is washed after antibody staining, the isotype control may not be different from an unstained cell. Second, the isotype is a different antibody, with different staining concentrations and possibly different brightness of the fluorophore (even if the same fluorophore is used). Therefore, while isotype controls are the established method of setting negative stained controls, they can introduce as many problems as they fix, if not more. Another issue that should considered with isotype controls is that one must be used for each fluorescence channel, compounding the number of reagents. Fluorescence detectors should be set with the same unstained or isotype controls, to ensure a large dynamic range, low signals for the controls, and proper gates for analysis. Regardless of whether an isotype or unstained control is used, the negative gate (i.e., cells without the antibody stain) will establish whether the samples analyzed from that point in the experiment are positive for the antigen of interest or not. Therefore, the utmost care should be taken when setting these control values.

The parameters set by the gates measure cell number and fluorescence or scatter intensity. The statistical values of interest for most flow measurements are fluorescence (typically median fluorescence intensity, MFI), scatter intensity, time of flight (pulse width of the scatter from each cell), and the coefficients of variation for each value. In many flow cytometer analyses, the cell number determined from fluorescence and scatter gates is the end target. In other studies, the fluorescence intensity is more important. For example, the kinetics of a fluorogenic reaction in a cell can be determined by MFI, and the cell count is essentially unimportant, provided enough cells are counted for accurate statistics.

Gating can be used to count cells with desired scatter or fluorescence signals to produce a final, analytical result. Gating can also be used, as mentioned above, to select a cell population before any analytical signal is measured. For example, for a typical, lysed blood sample, gating lymphocytes using forward and side scatter allows measuring of anti-CD4 staining to occur without monocyte interference. However, if gating is not used, then the anti-CD4 population contains lymphocytes, monocytes, and any other cell with autofluorescence or fluorescence crossover to be counted as well.

Most commercially available flow cytometers separate fluorescence into discrete bands, and intensity is detected for each channel. It is possible, however, to perform additional fluorescence-based measurements on cells in flow cytometry. In microscopy analysis of cells, fluorescence lifetime, anisotropy, and Förster resonance energy transfer (FRET) are readily implemented. Lifetime measurements in a stock flow cytometer would be difficult, as the excitation source must be pulsed or modulated. The latter is the more realistic approach, since most lasers used in flow cytometers are continuous-wave types. However, the typical dwell time of a cell in the laser beam is on the order of several tens of microseconds, which may not provide sufficient time for frequency-resolved lifetime measurements. Fluorescence anisotropy requires some modification to the optical detection system, and should not be undertaken unless one has permission from the instrument's owner, and has the optics experience to do so. Instruments modified to measure anisotropy split fluorescence from a given wavelength band according to polarization (with respect to the laser polarization). Investigations of free and bound probes, as well as studies of cell viscosity, are possible. FRET measurements require little or no modification of the instrument, provided that the excitation and emission wavelengths of the donor and acceptor match those of the instrument. In fact, tandem conjugates (e.g., PE-Cy5, APC-Cy7) use FRET to extend the wavelength range of most instruments.

6.8   EXAMPLE FLOW-CYTOMETRY ASSAYS

Chapter 9 covers many common cell assays, but in this section the assays that are particularly well suited to flow cytometry are discussed briefly. FRET measurements of fluorescent proteins, protein expression, viability, apoptosis, cell cycle, and intracellular antigens can all be measured, as well as a myriad of other assays. Flow cytometry measurements are particularly well suited to situations where high cell counts are needed, photobleaching is an issue, and morphology is unimportant. While there are hundreds of possible assays that can be performed by flow cytometry, three that showcase the detection power of flow cytometers are listed here.

6.8.1   Cell Cycle

Cell-cycle measurements require high cell counts and are particularly well suited to flow cytometry. This is evident when one considers that, to capture a view of the cell cycle from a population, a few cells measured on a microscope would not suffice. Fluorescence detection is linear, since cells vary only by a factor of two to four sets of chromosomes. Even in highly aneuploid cell lines, the difference between the maximum and minimum DNA content is small enough to use a linear scale. If the flow cytometer is equipped with a UV argon-ion laser (or other source) capable of exciting the Hoechst dyes, then direct measurement of DNA content can be accomplished without further cell handling. The Hoechst and 4',6-diamidino-2-phenylindole (DAPI) dyes are cell permeant and do not stain RNA at the same level as DNA. Using other dyes is also possible, but most other dyes are either cell permeant or RNA insensitive, but not both. Acridine orange is an exception to this rule of thumb, as RNA-and DNA-bound dye have different emission wavelengths and the dye is cell permeant. Cell-cycle assays based on impermeant dyes that are not DNA-selective require that the cell is permeabilized and that RNA is removed first by treating the cells with RNase. A simple protocol for cell cycles using these types of dyes is given in Protocol 9.11 (Chapter 9).

6.8.2   Fluorogenic Reactions

Fluorogenic reagents – those that remain nonfluorescent until a reaction occurs – are useful analytical probes, especially when one considers the reduced sample background and simplified assay protocols (e.g., Protocol 9.5, Table 9.7; Chapter 9). Examples of fluorogenic reagents include DNA dyes that increase quantum yield upon binding, bisamide rhodamine derivatives [9], acetoxymonoesters, and amine-staining reagents. One can, in principle, observe the generation of fluorescent product in a group of cells by microscopy. The major limitation to this approach – in addition to low cell counts – is that the cell is photobleached with each measurement of dye intensity vs. time. If the spatial location of the dye in the cell is unimportant, then production of a fluorogenic product in a cell sample can be analyzed by flow cytometry without photobleaching. It is important to note that with each measurement in time, different cells will be analyzed; however, the production of the dye in the total population can be ascertained as a function of time.

6.8.3   Measuring Antigen Density on Cell Surfaces

Cell-surface antigen density is a critical parameter in affinity cell separations (Chapter 5) and is readily measured by flow cytometry [10]. Given the variation in antigen density between cells of a given type, a large cell count is required for this type of analysis. Antigen density is calibrated using beads (microspheres) coated in three different approaches (Protocol 9.12, Chapter 9). First, beads with known antigen densities can be used and stained with the fluorophore-conjugated antibody. Second, beads with known densities of secondary antibodies can be used. For example, if mouse anti-human CD95 is used to assay CD95 density on human cells, then beads with anti-mouse antibodies can be used to calibrate the fluorescence response. A third, and simpler, approach is to use beads with known numbers of fluorophores attached to the surface [10]. This approach allows for greater flexibility, since only the fluorophore must be matched. However, in this approach, the valency of the antibody and number of fluorophores per antibody must be known.

6.9   NO-FLOW CYTOMETRY

Flow cytometry will continue to make great strides in cell analysis, but there is an inherent limitation to most flow instruments with regards to size, complexity, and cost. For most research and clinical settings, these factors are not significant barriers to using flow cytometry. However, when one considers some of the most pressing needs for flow cytometry measurements, namely CD4 T cell counting of patients with AIDS, the need for more accessible instruments becomes clear. It is not practical, for example, to establish a flow facility in remote locations. The cost, upkeep, and personnel required would be a major barrier to helping those who need medical care the most. One solution, proposed by Howard Shapiro [11], is to remove the flow aspect, and use wide-field imaging to measure cells. The concept requires low magnification to achieve high cell counts while operating at a spatial resolution that renders morphology irrelevant. The end result is a portable, low-cost cytometry platform that uses no flow, no moving parts, and could potentially be used in the future for resource-poor medical diagnostics.

The allure of this approach should, however, extend beyond the very important needs of resource poor medicine. In an ever-tightening funding climate, even researchers in the laboratory may find Shapiro's approach intriguing. Scatter measurements and sorting options are not possible in this format, but multi-color fluorescence cytometry is readily available. Many variations on the theme can exist, and a high cell count, low spatial resolution instrument can be set up quickly at a fraction of the cost of a flow cytometer. While flow instruments will always have an important place in cell analysis, this simple approach should also be investigated by those interested in cytometry.

6.10 CONCLUSION

Flow cytometry is perhaps one of the most powerful techniques for cell analysis. It is capable of sorting (Chapter 5), high-speed cell counts, and multi-parameter measurements. However, for all of the capabilities flow cytometry offers, it is a technique that is loaded with pitfalls if the user is not properly trained. Even with training, it can take some amount of time to adjust and develop a good "eye" for setting up the instrument and analyzing the data. In the analysis aspect of flow cytometry, there are many opportunities to introduce systematic or random bias that affect a measurement.

The cell sample, often the overlooked source of error, must be of the correct composition, size, and concentration for proper flow measurements. Also, as analysis becomes more complex (i.e., more parameters measured simultaneously), the need for proper controls and compensation increases. It is recommended that, whenever possible, the fewest parameters necessary should be measured to reduce experiment complexity. There is an unfortunate tendency to use every available fluorescence and scatter channel on an instrument, even when the cost and risk of error outweigh any benefit. An objective evaluation of the experiment will determine the correct number of parameters to measure from the cell sample.

Flow cytometry is a relatively mature technique, but each year new instruments that are either smaller (but unfortunately not much less expensive) or more powerful are introduced. While the cost of these instruments is high, it is worth obtaining one (either through individual or group ownership) when large cell counts or sorting are required.

REFERENCES

1. Shapiro, H.M. (2003) *Practical Flow Cytometry*, 4th edn, Wiley-Liss, New York.
2. Robinson J.P. (2008) *Handbook of Flow Cytometry Methods*, Wiley-Liss, New York.
3. Longobardi Givan, A. (2001) *Flow Cytometry: First Principles*, 2nd edn, Wiley-Liss, New York.
4. Watson, J.V. (2005) *Flow Cytometry Data Analysis: Basic Concepts and Statistics*, Cambridge Press.
5. Perfetto, S.P., Chattopadhyay, P.K., and Roederer, M. (2004) Seventeen-colour flor cytometry: unraveling the immune system. *Nature Reviews Immunology*, 4, 648–655.

6. Herzenberg, L.A., Tung, J., Moore, W.A. *et al.* (2006) Interpreting flow cytometry data: a guide for the perplexed. *Nature Immunology*, **7**, 681–685.

7. Habbersett, R.C. and Jett, J.H. (2004) An analytical system based on a compact flow cytometer for DNA fragment sizing and single-molecule detection. *Cytometry Part A*, **60**, 125–134.

8. Eckoff, R.K. (1969) A static investigation of the coulter principle of particle sizing. *Journal of Physics E*, **2**, 973–977.

9. Hug, H., Los, M., Hirt, W., and Debatin, K.-M. (1999) Rhodamine 110-linked amino acids and peptides as substrates to measure caspase activity upon apoptosis induction in intact cells. *Biochemistry*, **38**, 13906–13911.

10. Davis, K.A., Abrams, B., Iyer, S.B. *et al.* (1998) Determination of CD4 antigen density on cells: role of antibody valency, avidity, clones, and conjugation. *Cytometry*, **33**, 197–205.

11. Shapiro, H.M. and Perlmutter, N.G. (2006) Personal cytometers: slow flow or no flow? *Cytometry Part A*, **69**, 620–630.

7

Analyzing Cells with Microfluidic Devices

7.1 INTRODUCTION

Microfluidic devices, particularly those used in cell analysis, are a relatively new area of research when compared to flow cytometry or microscopy. However, the last 15 years have seen an explosion in both microfluidic technologies and their application to cells. Advances in microfluidics have enabled new geometries, flow regimes, and 3D cell devices to be fabricated. Along with progression of the cutting edge in microfluidics, the widespread availability of new, user-friendly fabrication approaches have brought microfluidic devices into laboratories that lack more advanced lithography facilities. This improved availability of microfluidic technologies has allowed cell researchers to adapt to newer technologies faster than before, and discover new methods of cell analysis. In addition to analytical devices, microfluidics have also been applied to the culture of cells. In many cases, the microfluidic chips mimic the function of traditional culture systems (flasks, bioreactors, etc.) albeit on a smaller scale. However, improvements in microfluidics have also initiated the development of newer culture techniques, co-cultures of complex geometries, and the ability to individually address cells or batches of cells. The more complex microfluidic systems are capable of cell culture, batch feeding, introduction of reagents, cell processing (lysis, etc.) and analysis on one chip. There is clearly a wide

Practical Cell Analysis   Dimitri Pappas
© 2010 John Wiley & Sons, Ltd

range of capabilities that microfluidics offer, and the increasing availability of such devices makes entering this field of research less difficult.

The use of microfluidics for cells can be grouped into three categories.
First, microfluidic chips can be used strictly for cell culture, or, second, for
analysis. Finally, the two can be combined for true "lab-on-a-chip"
functionality. Analyses conducted on chips include flow cytometry,
microscopy, cell separations, cell lysis and chemical analysis via electrophoresis, laser-induced fluorescence, or coupling to off-line systems such
as mass spectrometry. Discussion of flow cytometry is found in Chapter 6,
and while the larger instruments used commercially have more powerful
features, chip-based flow cytometry allows for smaller volumes and
customizable flow systems. Discussion of microfluidic cell separations is
found in Chapter 5, although the microfabrication aspects of this topic are
found in this chapter.

7.2   ADVANTAGES OF MICROFLUIDICS

Microfluidics are a relatively new area of research, and there is little
surprise that a great deal of attention is being paid to both the fundamental
and applied aspects of lab-on-a-chip devices. Microfluidics can mimic the
function of traditional culture on a smaller scale, or introduce capabilities
that would be either impossible or impractical for larger-scale systems.
The fluid volume of the device, for example, is typically much less than one
milliliter. The reduced volume of microfluidic culture and/or analysis
systems means that smaller volumes of medium and reagents are consumed and less waste generated. For example, if a chip-based culture is
going to be tested using an expensive or rare reagent, the smaller volume
allows for more tests. The reduced volume also allows for many different
cultures to be placed on the same device or set of devices. If a new
fluorescence substrate is to be tested and titrated, many cultures would be
required to cover the desired concentration range. If 16 concentrations
need to be tested, then 16 Petri dishes (10 ml each) or 10 flasks (25 ml each)
would be needed. One can, of course, also conduct the experiment in a
well plate (several hundred microliters at the maximum). However, if the
sample volume decreases, the lifetime of the cell population also decreases, since the medium volume reduces as well. Microfluidic devices
allow for sample volumes on the order of microwell plates or smaller, and
also enable medium exchange for long culture times (see Section 7.8 for
examples of adherent and suspended cell-culture systems). The same 16
experiments can be conducted using microfluidic culture systems that each

have sample volumes on the order of <100 µl, including medium flow into the device.

The scale of the microfluidic device is also important. Studies of mass transport to and from cell culture are difficult to conduct when the fluid volume above the culture surface is large. Studies of cell-to-cell communication, cell movement, and mass transport in and around cultures can all be conducted with relative ease in microfluidic systems. Cell–cell interactions, such as using flow channels to direct collision between cells, are also difficult to study in traditional fluid systems. Also, studies of cell adhesion [1], cell trapping [2], and novel flow profiles [3] are readily implemented in microchannels of varying geometries. Another interesting application of microfluidic devices is the study of cell crowding, where a culture occupies the entire available space in a microchannel. The study of mass transport in these 3D cultures can shed new light on tumor growth prior to angiogenesis, or to the expansion of bacterial biofilms.

One of the most interesting and promising areas of microfluidic cell analysis is the generation of gradients. There are several methods for generating a concentration gradient across a microfluidic chip [4,5], and the key advantage to doing so is that a large range of reagent concentrations can be tested on cells in a single device (Figure 7.1). The main benefit of gradient generation is that a single input sample (i.e., a single standard concentration) can be transformed into the desired gradient of concentrations. The cells on the device are exposed to the gradient for the desired time. An example of the utility of this approach would be the study of a cytotoxic compound. By careful generation of the compound's gradient, the cytotoxicity on the target cells could be tested. In this case, the spatial location of the cell can be used to identify the concentration of the test compound. Other examples include testing cells to determine the minimum amount of fetal serum needed to maintain a culture (Chapter 3), or to study a receptor-induced process as a function of concentration. While the method of generating gradients may change, the end result is that a culture or array of cultures can be addressed over more concentration values than is practical (or even possible) with conventional techniques.

Another reason why microfluidics and cells have had a successful partnering is that reagent manipulation and mixing can be accomplished in new ways to enable new cell analyses. For example, the laminar flow profiles of microfluidic devices allows reagent mixing and delivery that can be controlled both spatially and temporally [6]. Microfluidic devices can deliver reagents to a specific point [7] for a specific time, making microfluidic systems well suited to dose-dependent studies. The idea of modulating a reagent over cells would be a daunting one in most culture

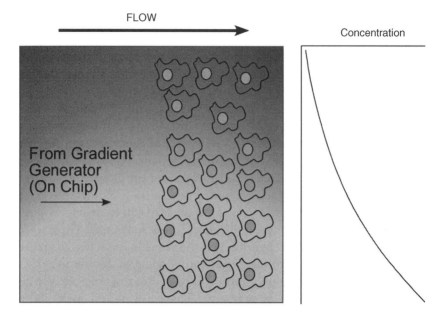

Figure 7.1   Use of gradients in microfluidic chips. Several different gradient generators can be used to produce a concentration gradient perpendicular to the fluid flow axis. One input concentration can then be used to produce a variety of concentrations of reagents. Toxicity, titration of reagents, and other dose–response studies can be conducted on a single cell sample over a large concentration range

formats, both because of the volume and the labor required. However, manipulation of microfluidic devices can occur on the chip itself or off of the chip in an automated fashion. It is possible to control more than one fluid stream, so that multiple reagents reach the target cells, either simultaneously or in a step-wise, temporal fashion, so that temporal dynamics can also be studied.

7.3   CONSIDERATIONS OF MICROFLUIDICS AND CELLS

One must consider the challenges of placing cells in the small volumes associated with microfluidic devices. The reduced volume that is often touted as a principle advantage to microfluidics requires constant fluid flow to supply fresh nutrients. Growing or keeping cells in a microfluidic device for more than an hour or so requires mass transport, otherwise the cells will die. Of course, the presence of fluid flow introduces another set of problems. For example, if cells were to be cultured in a simple microfluidic channel, there must be a time after the cells are introduced where no flow is present. This lag time allows cells to settle to the surface and begin to attach. Fluid flow must, however, resume after some time, otherwise the cells will die from lack of nutrients. Starting fluid flow too early will result in cells being flushed from the channel in question. This example would prove even more difficult for suspended cells, which would be removed from the channel unless anchored in some artificial manner. This problem has been solved in a number of ways [8–12], including micropillars to shield cultures, side-channels to reduce flow, and high-aspect-ratio culture chambers. These reduced-flow methods have allowed cells to be inoculated into the device and flow to begin immediately. The end result is an undisturbed culture (in some cases, suspended cells have been cultured) that still allows for mass transport of nutrients and waste.

When considering cell devices, the effect of gas exchange must also be considered. In glass and some plastic microfluidic chips, gas supply is controlled strictly by medium flow. In polydimethylsiloxane (PDMS) devices, however, the silicone material itself can exchange gases. This is often cited as a key advantage to PDMS devices; however, evaporation is also more likely to occur in PDMS chips than in those made entirely of glass. In a novel approach to culturing hepatocytes [13], a fluidic layer and adjacent chip layer for gas control were combined for precise control of the culture microenvironment. This example underscores one of the greatest strengths of microfluidics, the ability to control cell conditions locally. Another consideration when analyzing cells by microfluidic methods is that during culture on the device, the limited surface area can affect growth. Compared to traditional culture devices, the small growth area of microfluidic chips limits culture times to several days [8], primarily due to contact inhibition of growing cultures. For most experiments, culture times of this length are not needed, but the number of cells and the chip culture area should be considered in any experiment.

The major limitation of handling cells in microfluidic devices is the effect of shear stress on the cells (Figure 7.2). In any flowing stream, the shear force encountered by the cells can be large. Shear force can remove cells from the surface, or limit viability due to stress on the cells. Fluid shear can be estimated using Equation 7.1 [14] for square/rectangular channels, and Equation 7.2 [15] for circular channels.

$$\tau_w = \frac{6\mu Q}{h^2 w} \tag{7.1}$$

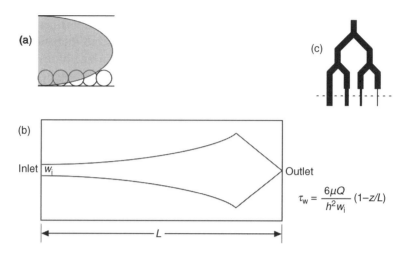

Figure 7.2 Concept of shear flow in a pressure-driven channel. Laminar flow (a) produces a steady-state shear (Equations 7.1 and 7.2) that acts on cells on the surface. A branched system with different channel cross-sections (b) can produce discrete shear forces at constant pressures (adapted from ref. [18]). A Hele–Shaw chamber (c) can be used to produce a linear shear force gradient along the fluid flow axis for adhesion studies (adapted from ref. [17])

$$\tau_{\mathrm{w}} = \frac{4\mu Q}{\pi r^3} \tag{7.2}$$

In Equations 7.1 and 7.2, μ is the fluid viscosity, Q is the volumetric flow rate, h and w are the channel height and width for square/rectangular channels, and r is the radius of the cylindrical channel. Given a microfluidic channel of $50 \times 50\,\mu\mathrm{m}$ (square), the shear force for a $1\,\mu\mathrm{l\,min}^{-1}$ flow rate can be calculated to be approximately $8\,\mathrm{dyn\,cm}^{-2}$ ($\mathrm{g\,cm}^{-1}\,\mathrm{s}^{-2}$), which is more than enough force to dislodge affinity-captured cells [13,16,17] in many cases. It is clear by Equations 7.1 and 7.2 that both the geometry and size of the chip, as well as the flow rate, must be considered when shear is to be minimized. In the cases where shear stress of cells is the desired parameter of measurement, the shear force can be varied by three different methods. One method, useful for existing cell-chip designs, is to simply vary the flow rate to adjust shear. A second method is to use a Hele–Shaw design ([17], Figure 7.2) to vary the shear force linearly along a chip for a given flow rate. The latter approach allows one to study a wide range of shear stress in a single experiment. A third approach [18] uses multiple channels that feature the same pressure drop and shear to evaluate the effects of flow on cell adhesion.

The geometric limitations of microfluidic cell analysis or culture devices depend in part on the methods of fabrication (see Section 7.4). However, some general rules apply to designing and using such devices for cell applications. First, as in flow cytometry, clogging of cells must be minimized. Clogging in many cases is worse than in a flow cytometer, because the flow rates are lower and the fluidic designs can become significantly more complex. The channel width and height must be optimized for the desired flow conditions, the correct shear force, and to minimize clogging of the chip. Even when hydrodynamic sheath flow is used, large cell clusters can foul an experiment. Another geometric limitation of cell-culture devices is that cell–cell collisions may occur due to the restricted volume of the chip. This effect may be desired in some cases, but in others, such as affinity separations of cells or cell docking, a colliding cell can dislodge one that was captured intentionally. In the case of cell-affinity chromatography, a careful balance must be made between a low ceiling to generate many cell–surface interactions but minimize cell–cell interactions. The device should also allow for expansion of the cell sample if the cells are to be kept viable and cultured for a period of time greater than a few hours.

Another consideration that should be made is the type of flow required in the device. Both external pumps and on-chip peristaltic devices (see Section 7.4, Figure 7.9) are capable of driving flow in the chip. The flow, however, is often laminar in these devices (Chapter 5), which can be problematic in some cases. Mixing is difficult unless pillars or other mechanical obstacles are added to disrupt laminae. Another issue with laminar flow is that cells – if they are traveling in the direction of the flow – will have different velocities, depending on their position in the flow channel. There are advantages to driving cells with laminar flow as well, such as the ability to hydrodynamically focus cells for flow cytometry. It is also possible to use electric fields to generate electro-osmotic flow. In this case, a microfluidic chip with a negatively charged channel surface (such as glass treated with basic solution to produce SiO^- functional groups) is subjected to an electric field. The charged glass surface produces a diffuse double layer of cations that drags the solution toward the negative terminal at a steady flow rate. Electro-osmotic flow can be generated for shorter periods of time than mechanical flow, and the electric fields used may lyse or adversely affect cells. However, electro-osmotic flow requires no moving parts, and the flow has a plug-like profile, where the flow velocities are the same across the flow channel.

7.4   OBTAINING MICROFLUIDIC CELL DEVICES

Microfluidic devices can either be obtained by fabrication or by buying pre-made chips commercially. Figure 7.3 lists some common microfabrication methods and materials, depending on application. Commercially available chips are currently limited to the designs that have found widespread acceptance. Simple channels, cross injection, and twin-T injection chips are readily available from several vendors (Figure 7.4 for example designs). However, most commercially available devices are made for chemical analysis (primarily electrophoretic separations). The current exception is the Bioanalyzer 2100 from Agilent. One must also check to ensure that the chip dimensions match the requirements for the cell application. For most culture applications, or deviation from premade chips, one must seek microfabrication facilities of some sort. Several in-depth discussions of microfabrication for fluidic applications have been published [19–21]. In this text, only a cursory discussion of these techniques is presented, as the major focus is cell-based applications. The setup of a microfabrication facility can be costly (when photolithographic methods are used). The cost of microfabrication will be dictated by the decision to use glass or polymeric materials, and by the required spatial resolution. If small (<20 μm) features for cell trapping, mixing, or other manipulation are required, the fabrication costs will be considerably

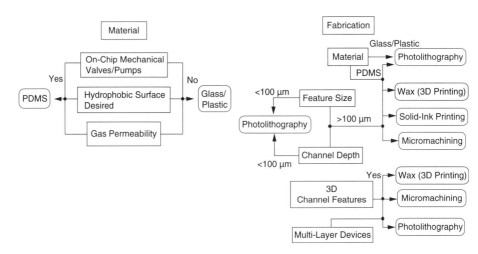

Figure 7.3   Flow chart of fabrication methods, based on both material and fabrication type. When choosing materials, mechanical deformation requires PDMS, while plastics and glass can be used when rigid structures are required. Fabrication choice will be dictated by material type, feature size, and 2- vs. 3D construction

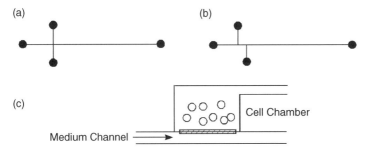

Figure 7.4   Three common, commercially available microfluidic designs. The cross (a) and split-T (b) style networks are useful for electrophoresis, cell lysis and electrophoresis, as well as some reagent mixing. The cross-flow system from the Onix culture system from CellAsic (c) allows cells to be cultured in low shear with adequate mass transport

higher. Similarly, if multi-layer devices featuring valves and pumps are needed, then additional instrumentation to align multiple layers may be required. In those cases, it may be best to find a collaborator with the existing infrastructure, or seek out a core microfluidic facility.

There are several options for in-house fabrication. At the higher end of capability, photolithography would offer the greatest flexibility and best spatial resolution (Figure 7.5). Briefly, a mask must first be generated, where the fluidic circuits for each layer are printed on a transmission substrate to cure the photoresist. The mask can be chrome-coated onto glass (the most expensive and precise method), or printed on high resolution (10 000–20 000 dpi) laser printer. Both mask options are typically made by third-party sources and range in price, based on complexity. A layer of photoresist (either positive or negative) such as SU-8 is coated to the desired thickness on a silicon wafer. Exposure to UV light through the mask then cures the photoresist, forming the basis of the microfluidic mold. If glass fabrication is used, then the photoresist (coated onto glass) serves to protect the surface and allows selective etching. For PDMS or plastic devices, the elastomer or uncured polymer is poured over the mold and cured against it. In PDMS chips, the aspect ratio of the channels/structures must be less than 10 : 1, otherwise support pillars will be needed to avoid collapse of the channel. The etched glass surface can be thermally bonded to another piece of glass to finish the device. PDMS layers can be sealed to glass or other PDMS layers via oxygen plasma, or sealed to each other by creating one layer with a high monomer–cross-linker ratio to a layer with a low ratio (for example [22] bonding of a 20 : 1 monomer–cross-linker ratio to a layer with a 5 : 1 ratio). For plastic chips, several bonding options exist. First, etched layers created by

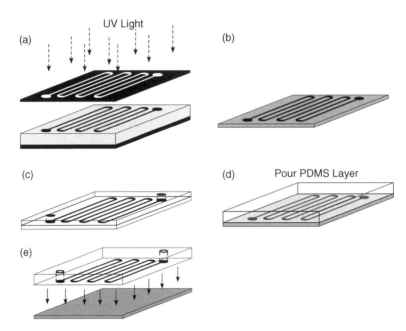

Figure 7.5 Photolithographic methods for microfluidic device fabrication. A mask is used to selectively irradiate a photoresist coated on a substrate (a), producing an image of the mask in the developed photoresist (b). For glass fabrication, the photoresist is used to shield the glass from etching, so that only the channels are etched. A glass top piece is hermetically sealed by thermal bonding, forming the all-glass device (c). For PDMS fabrication, a layer of PDMS is poured over the photoresist mold (d), creating the fluid network. The resulting PDMS fluid network is then sealed against a substrate (e), forming the PDMS device (the substrate can be glass, another piece of PDMS, or several different polymer types). In PDMS, the sealing can be irreversible or reversible, depending on need

photolithography can be thermally bonded in a manner similar to glass chips, albeit at different temperatures. The main problem with thermal bonding of plastic chips is possible deformation of the devices during the adhesion process. A simpler, room temperature approach is to solvent-weld the plastic layers using a bead of solvent to dissolve the two layers into each other. Once the solvent evaporates, a unified device is created. In order to solvent-weld chips without deforming (dissolving) the fluidic channels, a sacrificial layer must be present [23,24]. Sacrificial materials undergo a phase change from liquid to solid in order to fill the fluidic channel and protect the channels from dissolving (Figure 7.6). Solvent welding then takes place. After the solvent weld is finished, the sacrificial layer is converted into its liquid form and removed from the channel, leaving the original fluidic circuit intact. This approach is particularly

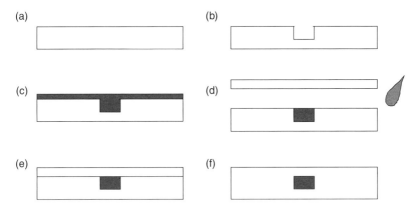

Figure 7.6   Fabrication of plastic chips by solvent bonding and sacrificial layers. A plastic piece (a) is etched, machined, or embossed with channels to make the fluid network (b). A sacrificial layer is added to protect the channel, and excess sacrifical material is removed after the solid phase change (c). A drop of solvent is added and a top plastic piece added (d), after which solvent welding fuses the two pieces (e). The sacrificial layer is melted and the final channel network is formed (f)

useful in multi-layer fluid circuits [24]. Sacrificial materials must be compatible with the plastic chip (i.e., should not permanently adhere, swell or in any way react with the polymer) and should undergo phase changes at reasonable temperatures. Several types of wax have been used, as well as water. The advantage of using a sacrificial layer is that solvent welding can be conducted at room temperature with minimal infrastructure. In addition, the roughness of the plastic surface is not as critical as in glass–glass bonding applications.

Since PDMS devices are unique in their fabrication, in that the chip material is molded against a master, there is tremendous flexibility in the type of mold-fabrication process. For glass, the chip must be etched directly, and for plastic direct etching or hot embossing are possible. In PDMS, however, virtually any material can serve as a mold. In addition to photolithographic processes mentioned above, three useful and accessible fabrication techniques have been used to generate a variety of devices. There are additional mold-making techniques reported in the literature, but these three examples highlight the differences in mold fabrication. Wax printing of various types has been used to make molds. 3D wax printers, which are used commercially to mold jewelry or prototype small parts, have been used [25] to produce complex 3D structures used to mold PDMS. The benefit of this approach is that fluid lines, interconnects, large well structures, and so on, can be fabricated simultaneously. The feature size is relatively large for many microfluidic applications (250 μm, [25]), but for cells, this fluid channel size is not prohibitive for many applications, such as flow cytometry or cell culture. 2D molds are also possible using a different type of wax printer [26]. Using a "solid ink" printer that is sold commercially for office printing needs, fluid traces can be printed on Mylar or transparency film. PDMS is then poured against the mold and peeled away after curing to fabricate the chip. The PDMS layer is then sealed to a PDMS slab or glass slide. The height of the PDMS channels depends on the height of the wax layer (20–30 µm is typical), and the smaller feature sizes depend on the resolution of the solid ink printer. Since the wax ink is heated and dispensed on the film, there are raster lines which can be smoothed out be briefly heating the wax, which reduces channel roughness.

Micromachining is also available, provided the local machine shop has tools of the right size. Using small end mills, channels can be produced in a variety of materials (metals, plastics, etc.) to produce molds for PDMS curing or embossing of plastics. Machining is inexpensive, but requires some time between device conception and final fabrication. In addition, the smaller end mills found in most machine shops will be limited to feature sizes greater than 100 µm. Another potential limitation of micromachining is that the round end mills used to make the mold will produce rounded channel intersections in some cases.

Another method for creating plastic chips is solvent etching. In this case, a PDMS mold of a fluidic circuit is generated (Figure 7.5). The PDMS stamp is then connected to inlet and outlet fluid connections and placed over a plastic [27]. A solvent capable of dissolving the plastic is then flowed through the PDMS channels, etching the same pattern into the plastic layer. The depth of the etch depends on solvent concentration, flow rate, and flow time (Figure 7.7). The key advantage of solvent-etched chips is the simple, straightforward nature of the fabrication process, once the PDMS stamp is made. The solvent used for etching should not swell or deform the PDMS, since the resulting fluidic circuit in the plastic piece will be distorted. The solvent etching method can be scaled to etch many chips simultaneously, potentially generating more chips per batch than other laboratory fabrication methods. Additional advantages of plastic chips include cost, experimental flexibility, and simpler secondary fabrication (drilling access holes, etc.).

PDMS or PDMS–glass chips offer similar flexibility, as well as the option to create multi-layer devices and devices with mechanical actuation on the chip. Multi-layer device fabrication with PDMS requires casting of thin films, each containing a portion of the 3D fluid system. Casting thin films can be accomplished by careful control of the

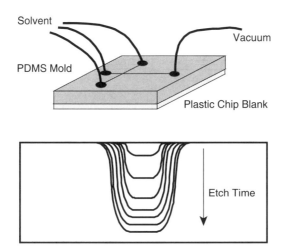

Figure 7.7   Solvent etching of glass and plastic chips. A solvent for glass or plastic is aspirated through channels in a PDMS mold (top). A replica of the fluid network is reproduced in the plastic or glass substrate (bottom), with etching time determining channel depth

spin-coating process, so only a thin layer of liquid PDMS mixture remains on the mold prior to curing. An alternative approach [28] uses a clamp system to produce reproducible, thin layers for 3D devices. Devices of this type create multi-layer, 3D networks that are bonded together to form a complex fluidic structure that is difficult to create in glass or plastic chips (Figure 7.8). From a cell culture or analysis standpoint, it is often necessary to move beyond traditional 2D chips, particularly when cell growth in three dimensions (e.g., biofilm growth) is desired. Other applications of 3D cell analysis chips include size-based sieves (where holes in the "floor" of a given layer sort cells to fluidic traces in a lower level) or microfluidic

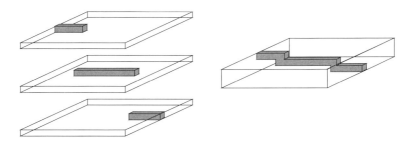

Figure 7.8   3D fluid networks. Using multiple layers of fluid traces, a 3D structure can be formed in a variety of geometries. Glass, plastic, and PDMS chips can all be made into 3D systems

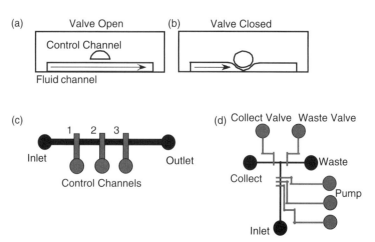

Figure 7.9 Active fluidic control in PDMS devices. Control channels (a) use compressed air to collapse fluid channels above or below (b). Several control channels can be placed in series and actuated sequentially, forming an on-chip peristaltic pump (c). Several control channels and on-chip peristaltic pumps can be combined into a functional device, such as the integrated fluorescence-activated cell sorter developed by Quake and colleagues (d, adapted from [31])

field flow fractionation techniques (See Chapter 5 for a discussion of field-flow cell separations).

Multi-layer PDMS devices can also be used beyond passive fluid flow to integrate mechanical actuation on the device (Figure 7.9). As shown in Figure 7.9, a control channel, placed above and orthogonal to a fluid channel can be pressurized with air or nitrogen gas to inflate the channel and close the adjacent fluid channel. The control channel can also be placed below the fluid channel, reducing the air pressure required to close a channel. The inflated control channel can completely restrict flow. PDMS valves are "normally open"-type switches. It is therefore difficult, if not impossible, to leave a device unpowered and keep fluid sections separate using the valves. Unpowered valves [29], based on screws embedded in a hard polymer layer, can be used to permanently close fluid channels for device transport, and so on. There is also a limit on how much fluid pressure can be regulated using air-pressure control channels.

Since control lines can be individually addressed, it is possible to use more than one valve in tandem. For example, one valve can be opened as another is closed. Computer-controlled solenoid valves can therefore be precisely timed to activate control lines at the correct time. Another application of multiple control lines is the on-chip peristaltic pump (Figure 7.9, [30]). The simplest PDMS peristaltic pump uses three control channels adjacent to each other. The three control channels activate and deactivate sequentially, pulling fluid plugs in one direction. While the off-chip hardware to achieve this type of pumping is not trivial, the ability to directly drive flow on the chip can be advantageous. For example, multiple pumps can be integrated on the same device. The flow rate of on chip pumps is pulsatile, which may affect separations, but other cell analyzes may not be affected. Also, the flow rate that can be achieved by on-chip pumps is lower than what is possible with off-chip pumps. However, the capabilities of on-chip valve and pumping arrangements allows cell devices to achieve complex fluid operations that were previously difficult or impossible.

One of the best examples of multiple control devices on the same chip is the integrated cell sorter developed by Quake and colleagues (Figure 7.9d, [31]). Quake's integrated cell-sorter chip contained a peristaltic pump to drive cells toward a T-shaped channel. At the channel junction, valves on each side arm operate out of phase so that only one channel allows fluid flow. Fluorescence detection occurs just before the junction, and the valves switch to pass a cell to waste or a collection vessel. Since the flow is generated on chip, it is possible to stop and reverse flow, allowing a cell to be measured more than once. Cells could therefore be analyzed rapidly, and when a target cell is detected, flow can be slowed, stopped, or even reversed to accommodate slower sorting mechanisms.

The material of choice for fabrication depends partly on the chip-making facilities as well as the application at hand. For example, UV-curable PDMS and epoxies can be made into chips with minimal fabrication infrastructure [32], but may not be amenable to certain analyses. If an outside or shared facility is used, then the existing methodologies will likely drive chip design. It is important to remember the nature of the most common chip materials when considering experiment design. PDMS, while simple to fabricate, is difficult to coat with metals for electrode manufacturing, such as separation by dielectrophoresis [33–35]. PDMS is also hydrophobic [36], therefore analysis of hydrophobic cell contents is difficult. For example, in an analysis of release of glycerol from adipocytes [37], glass chips offer lower adhesion of the target analyte, even if cell culture is less than ideal when compared to PDMS.

7.5   MICROFLUIDIC FLOW CYTOMETRY

The number of cell analysis methods performed in microfluidic devices is too large to chronicle in this text. Suffice to say, the expanding popularity of microfluidics is concurrent with new methodologies developed at the cellular level. From miniaturization of existing techniques, such as flow cytometry, to advanced methods of cell separation, processing, and culture, microfluidics have demonstrated tremendous advantages over other techniques.

There is little surprise that one of the first cell-based analyses conducted in microfluidic chips was flow cytometry [38–40]. Flow cytometry (Chapter 6) is one of the most widely used and information-rich analytical techniques for cells. There is also a large market for flow cytometry, flow sorting, and related products. Several companies have developed or are in the process of developing commercial microfluidic flow cytometers. However, a criticism of microfluidic flow cytometry – and many lab-on-a-chip techniques – is that the ancillary detection equipment is typically on the order of an inverted epifluorescent microscope in size and complexity. Even the Agilent 2100 bioanalyzer, which also has a cell chip for cytometry, occupies the same space as a small desktop computer. When striving for miniaturization of any technique, one must ask if the effort to create a microfluidic device outweighs the unique benefits microfluidics offer. For example, in the case of flow cytometry, does a small chip really outperform a bench-top instrument? For capillary electrophoresis separations, microfluidics have proven to allow multiplexed separations, or the integration of reaction vessels and separations in the same device. In flow cytometry, one could envision a chip where a raw sample was stained, fixed, and analyzed simultaneously. Most flow-cytometer chips demonstrated in the literature, however, do not offer these functionalities. Nevertheless, previous flow-cytometry chips have paved the way for future designs that may ultimately reduce labor costs, reagent volumes, and cross-contamination between runs.

Figure 7.10 shows two established approaches for on-chip flow cytometry. Both systems use hydrodynamic flow, but each varies in the execution of the flow profile. Agilent's CellChip for the 2100 Bioanalyzer (Figure 7.10c) uses an air pump to pull flow through six channels of identical length. The device reads each channel sequentially, and hydrodynamic flow pushes the cells along the edge of each channel. The CellChip is disposable, and does offer some advantages with regard to sample throughput. The main drawback is that the bioanalyzer air pump operates all six samples simultaneously, so as each channel is read, the subsequent samples are also being depleted. If cell concentrations are high, or the desired number of counts is low, then sample depletion is not as problematic. The second, and more common approach is to mimic the hydrodynamic flow arrangement of conventional flow cytometers, in a 2D

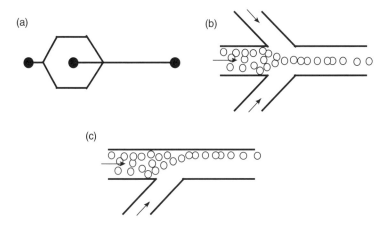

Figure 7.10 On-chip flow cytometry with hydrodynamic focusing. Two inlets can be used to merge sheath fluid and sample (a). In conventional on-chip focusing, two side channels introduce sheath fluid and cells are focused in the center of the stream (b). In the CellChip approach from Agilent, one sheath channel focuses cells to a wall (c)

fashion. Since the flow regime is restricted to two dimensions, the fluid flow rate does not currently match that of bench-top flow cytometers, extending analysis time. In addition, the smaller fluidic dimensions place greater restrictions on cell aggregates and clogging. Several companies have tried to develop flow instruments based on microfluidic architectures, although none have displaced conventional systems to date. There is, however, a great attraction to creating flow instruments using microfluidic devices. First, it is possible to integrate multiple sample steps into one device. It is possible to envision a system that combines staining, fixation, permeabilization, and labeling in one chip. The user would then simply aliquot the samples and run the analysis. The reduced risk of blood-borne pathogen exposure, and the reduced reagent consumption and labor costs would be an attractive advantage to these systems.

In addition to the ability to stain and fix samples, it is also possible to incorporate other cell separation techniques onto the device before the flow cytometer section of the chip. For example, whole blood can be loaded onto the chip and passed into a section of the chip for cell lysis of erythrocytes. The recovered leukocytes can then be passed on to the flow-cytometer section for measurement without the interference of the red blood cells. Another on-chip method to remove red blood cells involves a sieving method for size-based selection of leukocytes. It would also be possible to combine a cell-separation method based on cell-affinity chromatography (Chapter 5) to isolate a particular cell type before flow

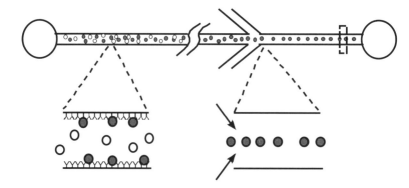

Figure 7.11   A proposed device for cell separation, flow cytometry, and detection. An on-chip cell affinity column (left) captures cells of interest; continuous loading is possible for rare-cell isolation. After unbound cells leave the chip, captured cells are eluted, hydrodynamically focused, and detected by flow cytometry on the same device (right)

cytometry detection. The key benefit of this approach would be that the cells can be enriched or depleted prior to detection. For example, an interfering cell type could be removed by the on-chip affinity column before the remaining cells are detected. Alternately, the target cell could be separated and enriched prior to staining and detection. Figure 7.11 depicts a potential device for cell separation, staining, and detection by flow cytometry. Fabrication in PDMS would allow for on-chip valves and pumps to be incorporated, although glass channels could be used if proper flow control was introduced. Cells enter the separation channel, coated with antibodies, aptamers, or other capture molecules. In the case of negative depletion, cells would have to be stained prior to introduction, so that all cells of interest can be detected (if fluorescence is used). If enrichment is desired, then the target cells can be captured in the column region, then eluted after all remaining cells in suspension are washed out of the capillary. In enrichment, one cell type can be separated, or a class of cells. For example, all T lymphocytes could be captured using an anti-CD3 column region, followed by staining to enumerate specific T cell types. Fluorescence staining could occur before separation or after (allowing only captured cells to be stained).

One of the benefits of on-chip flow cytometry is that detection typically occurs via a microscope. All of the possible functions of a microscope are therefore available for detection. Fluorescence, scatter (e.g., dark field), and polarization are available. Since most microscopes are more customizable than bench-top flow cytometers, it is relatively straightforward to add advanced fluorescence techniques such as lifetime, polarization (and anisotropy), and wavelength (e.g., spectral measurements of cells). Lasers or lamps can be used for excitation, and unique opportunities for imaging are also available.

As mentioned in Section 7.4, the microfluidic valve-sorting mechanism allows cells to be diverted between waste and collection wells. A different approach, developed by Dittrich and Schwille [41], uses electro-osmotic flow to divert cells between collection and waste streams. There are several other methods available for cell sorting to divert fluid flow and collect a cell type. When selecting the appropriate method, one must consider several key parameters. First, what is the desired sorting speed? Microfluidic devices are not likely to achieve sorting speeds that approach conventional systems in the near future. If a slower sorting speed (less than 1000 cells per minute) are acceptable, then microfluidic devices may meet the requirements. Other considerations of microfluidic-flow sorting instruments include closed (sterile) systems, cell size (for example, it is possible to sort very large cells or cell clusters since many fabrication techniques allow for customizable fluid dimensions), and the ability to couple the sorting mechanism to additional analyses. For example, it is possible to sort cells and then culture them on the same device.

7.6   CELL SEPARATIONS

Cell-separation methods are covered in detail in Chapter 5. While many formats exist for cell separations, microfluidic formats are particularly well suited and have gained in popularity in recent years. One of the most promising cell separation methods to be demonstrated in microfluidics is magnetic sorting (MACS). MACS separations are routinely performed by hand in the laboratory, but a flow-through method (See Figure 5.6, Chapter 5) allows for continuous enrichment. One example from Whiteside's group uses a simple magnet to deflect MACS-labeled cells into a collection channel while unlabeled cells pass to waste [42]. This is an excellent example of the simplicity of microfluidic MACS systems; they use commercially available, tested reagents in a different format. It is possible to envision a system where multiple magnets are located on the same chip, each of differing strength. It would then be possible to sort several cell types, each labeled with progressively smaller beads, so that the largest bead would be deflected by the weakest magnet, and so forth.

The simplest MACS application for microfluidics would be one based on traditional MACS experiments, but miniaturized to allow for continuous flow. A microfluidic device could be envisioned as a simple, straight channel. Chips of this type are available commercially, or could be prepared easily. The final chip would be placed on top of a magnet and the MACS-bead stained sample flowed through using a pump or hand injection. Washing with a buffer would remove nonspecifically bound cells, and the device would then be removed from the magnetic field to free the captured cells. This device could then be used for additional experiments, such as lysis and electrophoretic separation (particularly if a T-shaped channel were used) or for cell culture.

In addition to affinity separations and MACS/FACS sorting, microfluidic sieving and lysis devices are capable of separating cells in ways conventional, large-scale fluidics cannot. Cross-flow sieving devices [43,44] are capable of taking advantage of size differences in cells for separation. Lysis devices, which use selective hypotonic rupture (Chapter 5) to lyse red blood cells before restoring salinity to the recovered leukocytes, can remove red blood cells in a flowing device, allowing for continuous sample processing in an automated fashion.

In dielectrophoretic separations of cells, attractive and repulsive forces in an electric field cause cells of different properties to migrate to different regions of the field. Cells of different types must have a difference in size or dielectric properties for efficient separation. The separation of cells of different type is governed by the crossover frequency [33]:

$$f_0 = \frac{\sigma_s}{2\sqrt{2}} \pi d C_{mem}, \tag{7.3}$$

where σ_s is the buffer conductivity, d is the cell diameter, and C_{mem} is the cell membrane capacitance. Dielectrophoresis could be used to isolate bacteria from blood, or to separate cell types where differences in d and C_{mem} are sufficient for sorting. Currently, the main roadblock to widespread dielectrophoresis use is the fabrication process, which is not accessible as widely as other microfluidic techniques.

Capillary electrophoresis has also been used to separate intact cells, although primarily prokaryotes. As mentioned in Chapter 5, the electric field used to generate flow must be sufficient to generate separation, but must not lyse or affect the cells in a negative way. The mobility of the cells through the capillary will be based on both the electrical properties of the cell and the drag force of the cell as it passes through solution. It is possible to conduct the same experiment in a microfluidic chip, using a commercially available chip, as shown in Figure 7.4. Electrophoretic separation of cells or organelles [45–47] is possible, but will require some type of fluorescence stain to identify cells. However, electrophoretic separations are attractive in that a cytometer capable of some level of separation can be generated with no moving parts. These chips are directly amenable to microscope separation and detection with simple equipment

While one must avoid cell lysis in electric fields used to separate cells, it is desirable in some cases to lyse cells and analyze them by microfluidic chips. There is a great deal of interest in proteomic and metabolomic studies of single cells by microfluidics and capillary electrophoresis. The general concept is that since a single cell can be loaded onto a chip and lysed, subsequent separation and detection of the cell contents can yield a vast amount of information. The advantage to performing this type of cell analysis on a chip is that many experiments can be multiplexed and performed either in series or parallel to improve throughput of single-cell experiments. Single-cell lysis in microfluidic devices can occur using a preliminary stage for lysis, followed by a secondary stage for separation and laser-induced fluorescence detection [48].

7.7   ANALYSIS OF CELL PRODUCTS

The detection of cellular products is another important aspect of cell analysis that can be addressed by microfluidics. The secretion products of intact cells can be assayed in microfluidic formats using a variety of analytical techniques. Since microfluidic chips are capable of cell culture and single-cell manipulation, these devices have unique and powerful detection capabilities. When designing a device for culture and analysis, one must consider both the type of cell and what will be analyzed. The concentration of the analyte (secreted into surrounding medium or buffer), its composition, reactivity, or interference with the culture medium must be accounted for in order for the experiment to work (see Chapter 4 for fluorescence of culture media). For most analyses of cell products, the analyte concentration is the driving factor for separation and detection design. If many cells are cultured in a well, and their aggregate secretions analyzed, then the concentration may be sufficient for standard fluorescence techniques. For trace measurements, sensitive methods such as electrochemical or single-molecule detection may be required. It is possible to integrate preconcentration methods on the device as well.

Aside from concentration, the type of analysis is also critical. If fluorescence is to be used, does the secreted or lysed analyte fluoresce natively? If the molecule is indeed fluorescent on its own, then detection of

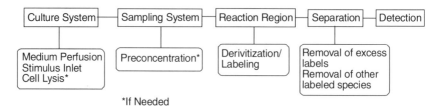

Figure 7.12   Components of a total culture and analysis system using microfluidics. A culture system maintains a cell sample, while a perfusion system allows for the introduction of stimulants. A sampling system removes cells or cellular products before any chemical reactions (such as derivitization for fluorescence) occur. A separation (chromatographic or electrophoretic) separates cellular products before detection (via microscope)

even a few copy numbers [49] are possible by confocal single-molecule fluorescence. If the molecule is not fluorescent, then there are several alternatives, including derivitization with a fluorescent conjugate or labeling with an antibody or other suitable capture molecule. Direct derivitization, where a covalent link is formed between the analyte and fluorophore, requires some type of separation step after the reaction to remove excess, unreacted fluorophore and also to remove any other species that are labeled alongside the analyte. Even if a fluorogenic derivitization agent is used, a separation step is required unless the reaction between the analyte and the agent is highly selective. These separation steps will likely have to be performed on the chip. It is possible to have a section of the chip (Figure 7.12) where culture occurs, a sampling fluidic system to move secreted or lysed products to the reaction zone, and a separation zone. The separation zone could be coupled with detection (e.g., fluorescence). If antibody or other capture molecules are used, then either some type of solid support (to anchor the capture molecule) or competitive assay must be implemented. In the case of a competitive assay [50], a fluorescent analyte is preloaded onto an antibody in solution. Displacement of the fluorescent analyte by the secreted one results in two fluorescent species being detected by electrophoretic separation – the antibody–label complex and the displaced label.

The block diagram shown in Figure 7.12 indicates some of the functionalities that must be considered for a total cell-analysis chip for cell product studies. The culture system, discussed in detail below, must be capable of homeostatic cell maintenance [51] during the experiment duration. The culture system must also be capable, in some cases, of introducing chemical stimuli so that a cell sample can be probed for a desired response. For example, in reports by Kennedy and co-workers [50], a single rat islet is placed in a well with fluid lines to introduce medium and a bolus of glucose. An electrophoresis channel samples the medium surrounding the islet for reaction and detection. The culture well is open to the laboratory to reduce shear stress.

The sampling system of the chip will depend largely on the experimental constraints and will likely involve some type of flow mechanism to transport analyte away from the culture region and to the reaction region. The flow rate of the sampling system will affect the sampling rate (time) and may affect the culture region as well. In the case of the latter, the flow rate must be kept below a shear value that would damage or negatively affect the cell sample. The flow rate must also be adjusted so that the analyte is not diluted significantly. For example, if the analyte is released in a relatively narrow burst, then the flow rate should be relatively fast to preserve that temporal resolution. However, if the analyte is secreted in a steady fashion, then a fast flow rate will dilute the sample.

The reaction system may also contain some type of preconcentration system. Preconcentration [52] is especially useful for slowly secreted compounds where temporal resolution is not as important. An on-chip preconcentrator reduces the burden on the detection system so that high-end optical detection is not needed. Preconcentration can occur by multiple methods, including ligand trapping (affinity separation), electrochemical stripping, and membrane or sol-gel trapping. The affinity approach requires an antibody or other capture molecule in the sampling or reaction system. Sol-gel or membrane trapping requires some level of affinity for the analyte in question (possibly based on hydrophobicity, for example). A continuous, low concentration of analyte from the cells is fed into the trapping region. The enriched analyte is later eluted or removed from the trapping region, further reacted (if required) and analyzed.

The reaction system is needed when the analyte does not have a native response for detection. For example, in most cases, protein secretion from cells cannot be detected by absorption, so the analyte must fluoresce for sensitive detection. Fluorescence in the deep-UV is possible, but the cell chip is likely to contain buffers containing proteins as well, thus generating high background fluorescence. Ideally, a reaction system would label or derivitize only the analyte. In practice, the labeling agent will also react with other compounds, requiring a separation in some cases. The exact labeling reagent will be based on the analyte itself. Protein and peptide analytes can take advantage of the wide range of amine-reactive dyes (see Chapter 9 for examples). The introduction of these reactive dyes requires that the reaction zone deliver the reagent at a precise point in time and space, and that adequate mixing occurs. Most microfluidic devices operate at low Reynolds number and therefore are laminar in nature. Laminar flow resists mixing, so serpentine channels, posts placed in the channel path, or other mixing strategies [53–55] must be incorporated to mix the reagent and sample. The labeled sample and excess reactive agent must then be separated for analysis. Fluorogenic reagents (Chapter 9) reduce the required separation resolution, as the excess, unreacted label is not detected. Whether fluorogenic substrates are used or not, this approach is useful for automated, continuous separation and detection of multiple analytes from a cell culture.

Another approach for on-chip detection of cellular products is to use an affinity-separation/concentration zone with a secondary antibody for detection. As shown in Figure 7.13, a region with immobilized antibodies (or other capture molecules) can preconcentrate the target analyte. After the preconcentration/sampling period, a second, fluorophore-conjugated antibody is used to label the analyte in a sandwich format. Detection occurs by imaging the region and integrating the fluorescence. It is also possible to forgo the second antibody and use a competitive assay. Detection is straightforward in either case using fluorescence imaging on a microscope, since no other separation step is needed, and the affinity capture is specific. However, the temporal resolution of this technique is poor, and it is a single-analyte method.

Most cell-culture/analysis systems will require some level of separation combined with detection. Optical detection is amenable to chip-based analysis and is already integrated into most microscopes. Since the

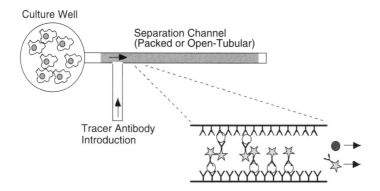

Figure 7.13 A microfluidic, affinity chromatography system for cellular products. A gentle sample stream carries cell products from the culture well to an on-chip separation channel. The channel is coated with ligands for the cellular products. On-chip detection can be achieved by adding a fluorescent tracer antibody, which binds to the captured cellular products and registers as a fluorescence readout

analytes from the cells are already in solution, the most straightforward approach to detection is to use either electrophoretic or chromatographic methods and optical detection at the end of the "column" or separation zone. Electrophoresis (combined with electro-osmotic flow) can be directly applied to most microchips – particularly glass ones – without additional fabrication. Electrode access is needed for the separation inlet and outlet. Injection can include electrokinetic injection in a twin-T or cross configuration (Figure 7.4) or flow-gated methods. One of the key considerations for coupling electrophoretic separations with cell cultures on the same device is that the buffers used and the applied electric field must not affect cell function and should be isolated from the culture region [50]. For chromatographic separations, either open-tubular or packed-bed columns can be generated on the chip for separation. Packed beds typically offer better chromatographic separations, but require higher pressures to operate. The practicality of driving a high-pressure flow on a microchip requires the other fluidic circuits to be isolated from the higher-pressure system, which is not always feasible. One option is to use large particles for the stationary phase support, so that the pressure can be generated by a standard pump used in chip operations. Detection can occur using either lasers or lamps, and microscopes outfitted with CCD or PMT detectors.

7.8   CELL CULTURE

While the cell-culture aspect of an integrated cell-analysis chip is outlined above, a detailed discussion of cell-culture chips – whether analytical capabilities are incorporated or not – is discussed here. As mentioned earlier in the chapter, there are several fundamental differences between traditional cell culture and cultures grown in chips. Flow profiles, mass transport, nutrient volume, and so on, require rational design of a culture chip. Culture chips can range from simple devices designed to support a group of cells [8] to complex systems for cell culture, processing, and analysis [22,49]. In this section, several parameters of culture systems are described, as well as a simple culture chip that can be made in the lab with minimal fabrication infrastructure.

The geometries of culture designs vary widely, in part because of different experimental needs. One must arrange cells in a well or chamber of a given geometry or volume, and provide access to the culture so that nutrients can be introduced and waste removed. Gas exchange is also critical, which is one of the reasons why PDMS chips tend to be used more often in long-term cultures. Multiple cultures can be conducted on the same device; however, each system should have its own fluid (and gas, if applicable) lines so that cultures do not cross contaminate. Another issue that needs to be considered with culture chips is that sterilization is not as straightforward, since the fluidic features may be too small for hydrogen peroxide or autoclave sterilization. Autoclaving will work in some cases, provided the chip and related interconnects can withstand the sterilization process. Flushing the system with 70% ethanol, followed by UV sterilization, offers rapid and effective sterilization, provided the ethanol solution can be removed from all of the chip volume.

If cells are exposed to flow, then some sort of anchoring step must be introduced (see Chapter 4) in order to keep cells in place. Anchoring can be achieved by coating the cell culture surfaces with poly-L-lysine, or fibronectin. In the case of low-shear designs [8–10], suspended cells can be grown without anchoring, since the flow does not directly interact with the culture. In microfluidic cell cultures that do not shield the cells from direct flow (and shear), then an attachment period – in the absence of flow – must be implemented to ensure cells are not flushed from the device. This attachment period must be short enough so that cells do not deplete the limited nutrients in the culture volume, but long enough to allow sufficient anchorage before flow is initiated. Since shear (see above) can negatively impact viability, cell structure, and cell function, flow-shielding designs should be employed or the flow rate should be low enough to allow medium perfusion without high shear stress. One way to lower shear without shielding is to increase the height of the culture chamber relative to the rest of the fluid lines. A taller culture chamber will allow for a longer anchoring time as well, since the medium volume increases. However, if the culture is to be manipulated by flow (e.g., removed from the culture region for on-chip cytometry measurements at a later point), then flow-shielding designs cannot be implemented.

When optimizing the flow rate for cultures on a chip, one must balance shear stress with nutrient replacement. In addition to viability measurements, both cell morphology and the degree of apoptosis should be assayed on the chip using fluorescence imaging (see Chapter 9). Shear stress may influence adherent cell morphology or alignment, as well as cell size. Apoptosis and viability will give an indication of cell function. By conducting a series of these assays as a function of flow rate (at the desired culture duration), then optimum flow rates can be determined for future devices. It is important to note that the variable shear and flow devices shown in Figure 7.2 would be particularly useful in these optimization experiments, as the experimental throughput is high.

An alternative approach to cell culture that eliminates many flow-related problems is to use a system such as the Onix cell culture system developed by CellAsic. In this approach (Figure 7.4), each culture chamber has a channel for cell introduction and removal, as well as an adjacent medium channel. The culture chamber and medium channel are joined by a microfilter wall that allows mass transport of fluid but keeps cells in the culture chamber. This device allows rapid changing of media and loading of reagents. The shear stress in this type of device is reduced as well. Each device has several culture chambers and integrates the microfluidic chip with a microwell plate for easier fluid handling.

The duration of the culture depends largely on the experiment at hand and the robustness of the cells. Some primary cells, for example, have a short culture time span even in traditional culture systems. However, for immortalized cells, the culture duration will depend on the culture chamber size, how well the chip perfuses medium, and also how well the system resists contamination. Any culture chamber, Petri dish, or flask, attempts to shield the cells inside from the ubiquitous threat of bacterial and fungal contamination (see Chapter 3). Unlike flasks, culture chips tend to have more opportunities for contamination due to chip leakage, fluid interconnects, and so on. The culture, if kept free of contamination, can continue until the cells occupy all available space or cells become senescent. Culture times of several days are possible if the system remains closed and proper incubation occurs. Incubation at the correct temperature requires that the device either remains in a thermostatted incubator between analyses or on a stage heater (Chapter 4) for the duration of the experiment. Which approach is chosen depends on whether the cells will be assayed continuously or at specified intervals. For experiments lasting multiple days, it is unlikely that continuous measurement will occur. In those cases, it would be simpler to leave the culture system in an incubator between measurements.

One of the biggest advantages of microfluidic analyses is that the chips can be customized for any experiment, provided one has the means to produce or buy the chips. Making the devices in one's own laboratory allows for faster turnaround from idea to final device, and for small variations in chip design to be tested rapidly. The following protocol will produce on-chip culture capabilities for PDMS-based devices (an option for all-glass devices is given as well). The culture chamber in question is but one of many equally useful options, but serves as an example that can be integrated into a variety of other microfluidic analyses.

PROTOCOL 7.1: LOW-SHEAR CELL-CULTURE CHIP

1. *Fabrication of Device Master and Microfluidic Components.* The choice of master fabrication depends in part on the facilities in place. Photolithographic methods will provide greater flexibility (see Figure 7.3) but will cost more than solid-ink or machined masters. The outline for this particular culture device is shown in Figure 7.14. The dimensions can be scaled depending on the needs of the experiment, but, for reference, a medium channel of 100 μm and a culture chamber of 350 μm will be used, and a sampling channel of 50 μm. The short channel joining the medium channel and culture chamber is also 50 μm. The medium channel operates at a flow rate that produces enough vortex flow to allow culture with high viability [8,56]. The sampling channel is smaller so that the medium flow is directed through the main channel primarily. One could also add a control layer with air lines to seal off the sampling channel, but in the approach shown in Figure 7.14, a small, continuous flow out of the culture chamber will provide constant sampling. The connecting channel length could range from 50–150 μm (in PDMS the aspect-ratio rule dictates that the channel height be at least 15 μm). The same device can be made of glass or plastic, although gas exchange will be more problematic than if PDMS is used. To access the fluid channels, holes can be punched in the PDMS or drilled into the glass or plastic. The bottom substrate should be glass in this application, particularly if microscopic investigation is desired. For

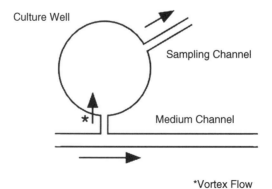

Figure 7.14 Simple, low-shear culture system based on vortex flow. A culture well is positioned off a medium channel (driven by a mechanical pump). Flow into the culture well produces a vortex at the channel intersection. A sampling channel can be added to observe culture products (see Figure 7.13)

devices made with plastic portions for the channels, a plastic bottom plate should be used.

It is also possible to "punch" a larger culture chamber out of PDMS so that a hole cutter is used to create a cylindrical culture well. The diameter of the culture well is dictated by the diameter of the cutting tool, and the well height is determined by the thickness of the PDMS layer. This approach creates an open well that must then be sealed with a small "roof" of PDMS to create a closed system. A hole cutter can be made using a piece of steel or copper tubing that is ground sharp. A larger well can be made in glass chips if a hole is drilled to create the culture chamber, and then a small "roof" of PDMS is attached to cover the hole and provide gas exchange (Section 7.3 of this protocol).

2. *Associated Hardware.* To drive flow through the main channel, a syringe or peristaltic pump capable of meeting the desired volumetric flow rate is attached to the chip. An on-chip pump is possible in PDMS devices, but for simplicity, off-chip pumps will be used in this protocol. Teflon or silicone tubing can be used to enhance oxygenation of the medium. If the device is made from PDMS, then steel tubing can be inserted in the punched holes to facilitate tubing connections. For glass chips, small plastic tubes can be glued to the device to allow tubing to be attached. The pump will drive fluid through the medium channel and finally to waste. The waste channel has several options, depending on the experiment. A simple well can be added to the device, where an open-ended glass or plastic cylinder is glued to the top of the chip and an access hole allows medium to fill the cylinder. This approach invites contamination – not likely to be a factor for short-term experiments – and requires the waste to be housed directly on the chip. Another approach is to add a tubing line at the distal end of the medium channel and have a small centrifuge tube serve as a waste receptacle. The centrifuge tube (0.5–1.0 ml size) can be open to the lab air, or sealed. In the case of sealed operation, a second tube connected to a 0.22 µm disk filter will prevent pressure buildup and contamination.

3. *Device Assembly.* After any fluid access holes are punched or drilled, the device must be assembled before tubing and other external connections are made. The device can be bonded to the glass substrate using an oxygen plasma (or solvent welding for plastic chips). For those without access to a plasma cleaner, a home-built system can be made with a residential microwave, vacuum pump,

and dessicator or other glass chamber [57,58]. Once the fluidic layer is sealed to the glass, any additional PDMS layers (such as the "roofs" outlined in Section 7.1), can be sealed to the fluidic layer. Since control (air) layers are omitted, alignment is not critical. The plasma exposure will depend on the plasma cleaner, but in a home-built microwave system with 1–5 torr of oxygen, a one-minute exposure will result in strong glass–PDMS or PDMS–PDMS seals. The assembly concludes by connecting the tubing to the supply and waste lines. It is best, if the device is to be used at a later date, to seal the two ends using Luer-locks and caps. Sterilization is typically the next step.

4. *Cell Inoculation.* These steps should be performed after steriliza-tion and in a laminar flow biosafety hood (See Chapter 2). Once all sterilization agent is removed, a sterile solution of poly-L-lysine, fibronectin, or serum can be injected into the culture well to promote cell attachment. The device, in particular the culture well, should then be filled with culture medium. The cell sample should be prepared for the appropriate cell concentration in the culture medium that will be used in the device. Injecting the cells in this particular cell-culture design can be done in two different methods. For PDMS chips, or glass devices with PDMS roof structures, a small-bore needle can be used to inject cells directly into the culture chamber. This direct approach is simple and rapid, but the small hole punctured in the PDMS may leak. A small drop of PDMS prepolymer and curing agent mixture can be placed on the injection site. This liquid PDMS can then cure at room temperature or in a 37 °C incubator and prevent leakage or contamination. In a PDMS device, a second approach is to manually compress the medium channel after the intersection with the well, and inject the cell suspension from the proximal end of the medium channel. The culture will then flow into the medium well. This operation is best conducted while observing the culture chamber on the microscope so that cells are not injected past the chamber. Once injection is complete, the com-pression on the medium channel is released. Since this device is a low-shear system, flow of culture medium through the medium channel can begin immediately without disturbing cells in the culture chamber [8].

5. *Culture Monitoring.* Operating this device using either a stage heater or other methods discussed in Chapter 4, one can observe the culture attachment and spreading in real time. If this is the first time this or

other culture chambers are used on a chip, then screening for apoptosis or cell proliferation (e.g., CD71 expression, see Chapter 9) will also aid in assessing the design's utility for cell culture.

7.9 CONCLUSION

The culture device outlined in Protocol 7.1 is just one example of relatively straightforward fluid circuits that can be added to chips for cell analysis. The flexibility in design, ability to combine multiple operations in a single chip, and the ability to manipulate cells in new ways will continue to ensure that microfluidic devices play an important role in cell analysis. The variety of fabrication methods and materials also allows a wide range of experiments to be developed and implemented for cell analysis. Other advantages of using microfluidic chips for cell analysis include the possibility to reduce contamination and/or biohazard risk, and the ability to couple microscopic investigation during the experiment operation.

A barrier to fabrication access remains for many investigators. PDMS devices, and non-lithographic methods such as solid ink printing, enable simple circuits to be designed; however for more complex, 3D devices or small feature sizes, photolithography remains the standard method. The formation of microfluidic fabrication facilities that share resources is a step in the positive direction to enable more researchers to put their ideas onto chips. While many microfluidic devices have focused on chemical separations, cell applications will continue to expand and enable new analytical techniques to be developed that cannot be realized in other formats.

REFERENCES

1. Wang, K., Solis-Wever, X., Aguas, C. *et al.* (2009) Differential mobility cytometry. *Analytical Chemistry*, **81**, 3334–3343.
2. Valero, A., Merino, F., Wolbers, F. *et al.* (2005) Apoptotic cell death dynamics of HL60 cells studied using a microfluidic cell trap device. *Lab on a Chip*, **5**, 49–55.
3. Shelby, J.P., Lim, D.S.W., Kuo, J.S., and Chiu, D.T. (2003) Microfluidic systems: high radial acceleration in microvortices. *Nature*, **425**, 38.
4. Irimia, D., Geba, D.A., and Toner, M. (2006) Universal microfluidic gradient generator. *Analytical Chemistry*, **78**, 3472–3477.
5. Vahey, M.D. and Voldman, J. (2009) High-throughput cell and particle characterization using isodielectric separation. *Analytical Chemistry*, **81**, 2446–2455.

6. Dasgupta, P.K., Surowiec, K., and Berg, J. (2002) Flow of multiple fluids in a small dimension. *Analtyical Chemistry*, **74**, 208A–213.
7. Wheeler, A.R., Throndset, W.R., Whelan, R.G. *et al.* (2003) Microfluidic device for single-cell analysis. *Analytical Chemistry*, **75**, 3581–3586.
8. Liu, K., Dang, D., Harrington, T. *et al.* (2008) Cell culture chip with low-shear mass transport. *Langmuir*, **24**, 5955–5960.
9. Kim, L., Toh, Y.-C., Voldman, J., and Yu, H. (2007) A practical guide to microfluidic perfusion culture of adherent mammalian cells. *Lab Chip*, **7**, 681–694.
10. Kim, L., Vahey, M.D., Lee, H.Y., and Voldman, J. (2006) Microfluidic arrays for logarithmically perfused embryonic stem cell culture. *Lab Chip*, **6**, 394–406.
11. Khadamhosseini, A., Yeh, J., Eng, G. *et al.* (2005) Cell docking inside microwells within reversibly sealed microfluidic channels for fabricating multiphenotype cell arrays. *Lab Chip*, **5**, 1380–1386.
12. Lee, P.J., Hung, P.J., Rao, V.M., and Lee, L.P. (2006) Nanoliter scale microbioreactor array for quantitative cell biology. *Biotechnology and Bioengineering*, **94**, 5–14.
13. Kane, B.J., Zinner, M.J., Yarmish, M.L., and Toner, M. (2006) Liver-specific functional studies in a microfluidic array of primary mammalian hepatocytes. *Analytical Chemistry*, **78**, 4291–4286.
14. Murthy, S.K., Sin, A., Tompkins, R.G., and Toner, M. (2004) Effects of flow and surface conditions on human lymphocyte isolation using microfluidic chambers. *Langmuir*, **20**, 11649–11655.
15. Reneman, R.S., Arts, T., and Hoeks, A.P.G. (2006) Wall shear stress—an important determinant of endothelial cell function and structure—in the arterial system in vivo. *Journal of Vascular Research*, **43**, 251–269.
16. Wang, K., Marshall, M.K., Garza, G., and Pappas, D. (2008) Open-tubular capillary cell affinity chromatography: single and tandem blood cell separation. *Analytical Chemistry*, **80**, 2118–2124.
17. Cheng, X., Irimia, D., Dixon, M. *et al.* (2007) A microfluidic device for practical label-free CD4 + T cell counting of HIV-infected subjects. *Lab on a Chip*, **7**, 170–178.
18. Lu, H., Koo, L.Y., Wang, W.M. *et al.* (2004) Microfluidic shear devices for quantitative analysis of cell adhesion. *Analytical Chemistry*, **76**, 5257–5264.
19. Kim, P., Kwon, K.W., Park, M.C. *et al.* (2008) Soft lithography for microfluidics: a review. *Biochip Journal*, **2**, 1–11.
20. Beebe, D.J., Mensing, G.A., and Walker, G.M. (2002) Physics and applications of microfluidics in biology. *Annual Review of Biomedical Engineering*, **4**, 261–286.
21. Abgrall, P. and Gue, A.M. (2007) Lab-on-chip technologies: making a microfluidic network and coupling it into a complete microsystem—a review. *Journal of Micromechanics and Microengineering*, **17**, R15–R49.
22. Gomez-Sjöberg, R., Leyrat, A.A., Pirone, D.M. *et al.* (2007) Versatile, fully automated, microfluidic cell culture system. *Analytical Chemistry*, **79**, 8557–8563, See Supporting Information for Microfabrication Protocol.
23. Koesdjojo, M.T., Tennico, Y.H., and Remcho, V.T. (2008) Fabrication of a microfluidic system for capillary electrophoresis using a two-stage embossing technique and solvent welding on poly(methyl methacrylate) with water as a sacrificial layer. *Analytical Chemistry*, **80**, 2311–2318.
24. Fuentes, H.V. and Woolley, A.T. (2008) Phase-changing sacrificial layer fabrication of multilayer polymer microfluidic devices. *Analytical Chemistry*, **80**, 333–339.

25. McDonald, J.C., Chabinyc, M.L., Metallo, S.J. *et al.* (2002) Prototyping of micro-fluidic devices in poly(dimethylsiloxane) using solid-object printing. *Analytical Chemistry*, **74**, 1537–1545.

26. Kaigala, G.V., Ho, S., Penterman, R., and Backhouse, C. (2007) Rapid prototyping of microfluidic devices with a wax printer. *Lab Chip*, **7**, 384–387.

27. Brister, P.C. and Weston, K.D. (2005) Patterned solvent delivery and etching for the fabrication of plastic microfluidic devices. *Analytical Chemistry*, **77**, 7478–7482.

28. Jo, B.-H., Van Lerberghe, L.M., Motsegood, K.M., and Beebe, D.J. (2000) Three-dimensional micro-channel fabrication in polydimethylsiloxane (PDMS) elastomer. *Journal of Microelectromechanical Systems*, **9**, 76–81.

29. Weibel, D.B., Kruithof, M., Potenta, S. *et al.* (2005) Torque-actuated valves for microfluidics. *Analytical Chemistry*, **77**, 4726–4733.

30. Cellar, N.C., Burns, S.T., Meiners, J.-C. *et al.* (2005) Microfluidic chip for low-flow push-pull perfusion sampling in vivo with on-line analysis of amino acids. *Analytical Chemistry*, **77**, 7067–7073.

31. Fu, A.Y., Chou, H.-P., Spense, C. *et al.* (2002) An integrated microfabricated cell sorter. *Analytical Chemistry*, **74**, 2451–2457.

32. Tsougeni, K., Tserepi, A., and Gogolides, E. (2007) Photosensitive poly(dimethylsiloxane) materials for microfluidic applications. *Microelectronic Engineering*, **84**, 1104–1108.

33. Das, C.M., Becker, F., Vernon, S. *et al.* (2005) Dielectrophoretic segregation of different human cell types on microscope slides. *Analytical Chemistry*, **77**, 2708–2719.

34. Gadish, N. and Voldman, J. (2006) High-throuput positive-dielectrophoresis bio-particle microconcentrator. *Analytical Chemistry*, **78**, 7870–7876.

35. Voldman, J., Gray, M.L., Toner, M., and Schmidt, M.A. (2002) A microfabrication-based dynamic array cytometer. *Analytical Chemistry*, **74**, 3984–3990.

36. Liu, K., Tian, Y., Pitchimani, R. *et al.* (2009) Characterization of PDMS-modified glass from cast-and-peel fabrication. *Talanta*, **79**, 333–338.

37. Clark, A.M., Sousa, K.M., Jennings, C. *et al.* (2009) Continuous-flow enzyme assay on a microfluidic chip for monitoring glycerol secretion from cultured adipocytes. *Analytical Chemistry*, **81**, 2350–2356.

38. Schrum, D.P., Culbertson, C.T., Jacobson, S.C., and Ramsey, J.M. (1999) Micro-chip flow cytometry using electrokinetic focusing. *Analytical Chemistry*, **71**, 4173–4177.

39. Yi, C., Li, C.-W., Ji, S., and Yang, M. (2006) Microfluidics technology for manipulation and analysis of biological cells. *Analytica Chimica Acta*, **560**, 1–23.

40. Pappas, D. and Wang, K. (2007) Cellular separations: a review of new challenges in analytical chemistry. *Analytica Chimica Acta*, **601**, 26–35.

41. Dittrich, P.S. and Schwille, P. (2003) An integrated microfluidic system for reaction, high-sensitivity detection, and sorting of fluorescent cells and particles. *Analytical Chemistry*, **75**, 5767–5774.

42. Xia, N., Hunt, T.P., Mayers, B.T. *et al.* (2006) Combined microfluidic-micromagnetic separation of living cells in continuous flow. *Biomedical Microdevices*, **8**, 299–308.

43. Murthy, S.K., Sethu, P., Vunjak-Novakovic, G. *et al.* (2006) Size-bases microfluidic enrichment of neonatal rat cardiac cell populations. *Biomedical Microdevices*, **8**, 231–237.

44. VanDelinder, V. and Groisman, A. (2007) Perfusion in microfluidic cross-flow: separation of white blood cells from whole blood and exchange of medium in a continuous flow. *Analytical Chemistry*, **79**, 2023–2030.

45. Kostal, V. and Arriaga, E.A. (2008) Recent advances in the analysis of biological particles by capillary electrophoresis. *Electrophoresis*, **29**, 2578–2586.

46. Chen, Y., Xiong, G., and Arriaga, E.A. (2007) CE analysis of the acidic organelles of a single cell. *Electrophoresis*, **28**, 2406–2415.

47. Chen, Y. and Arriaga, E.A. (2007) Individual electrophoretic mobilities of liposomes and acidic organelles displaying pH gradients across their membranes. *Langmuir*, **23**, 5584–5590.

48. McClain, M.A., Culbertson, C.T., Jacobson, S.C. *et al.* (2003) Microfluidic devices for the high-throughput chemical analysis of cells. *Analytical Chemistry*, **75**, 5646–5655.

49. Huang, B., Wu, H., Bhaya, D. *et al.* (2007) Counting low-copy number proteins in a single cell. *Science*, **315**, 81–84.

50. Dishinger, J.F. and Kennedy, R.T. (2007) Serial immunoassays in parallel on a microfluidic chip for monitoring hormone secretion from living cells. *Analytical Chemistry*, **79**, 947–954.

51. Reif, R.D., Martinez, M.M., Wang, K., and Pappas, D. (2009) Simultaneous cell capture and induction of apoptosis using an anti-CD95 affinity microdevice. *Analytical and Bioanalytical Chemistry*, **395**, 787–795.

52. Foote, R.S., Khandurina, J., Jacobson, S.C., and Ramsey, J.M. (2005) Preconcentration of proteins on microfluidic devices using porous silica membranes. *Analytical Chemistry*, **77**, 57–63.

53. Oberti, S., Neild, A., and Ng, T.W. (2009) Microfluidic mixing under low frequency vibration. *Lab Chip*, **9**, 1435–1438.

54. Lee, S.H., van Noort, D., Lee, J.Y. *et al.* (2009) Effective mixing in a microfluidic chip using magentic particles. *Lab Chip*, **9**, 479–482.

55. Glasgow, I., Batton, J., and Aubry, N. (2004) Electroosmotic mixing in microchannels. *Lab Chip*, **6**, 558–562.

56. Liu, K., Tian, Y., Burrows, S.M. *et al.* (2009) Mapping vortex-like hydrodynamic flow in microfluidic networks using fluorescence correlation spectroscopy. *Analytica Chimica Acta*, **651**, 85–90.

57. Ginn, B.T. and Steinbock, O. (2003) Polymer surface modification using microwave-oven-generated plasma. *Langmuir*, **19**, 8117–8118.

58. Hui, A.Y.N., Wang, G., Lin, B., and Chan, W.-T. (2005) Microwave plasma treatment of polymer surface for irreversible sealing of microfluidic devices. *Lab Chip*, **5**, 1173–1177.

8

Statistical Considerations

8.1 INTRODUCTION

In any scientific measurement or result, there is always a value, datum, or figure of merit. Along with this value, however, there must be proper statistical reporting. While errors and statistics are often treated as an afterthought, they are fundamental to the experimental process. Failure to address statistics often results in inconclusive data in the best case, and misleading data in the worst. While statistics are routinely ignored or abused for political or marketing campaigns, cell analysis requires that attention is paid to statistical treatment.

To demonstrate the importance of statistical analysis in cell measurements, it is important to consider what is truly being measured. Presumably there is a parameter of interest, such as cell size, or fluorescence intensity, that is going to be measured in a cell sample. That cell sample may be representative of a larger population. Within the cell sample there will be biological and chemical variation that arises from the randomness of nature. In addition to that variability (an important parameter that should be determined in its own right), the instrumental approach will impart an error that will convolute with the natural variation of the sample. Figure 8.1 illustrates some of the errors that can affect many cell analyses.

In his landmark work on cell counting, W. Sealy Gosset outlined the statistics and error of counting cells on a hemacytometer [1]. Gosset, identified as "Student" on the manuscript and equations that bear the same name, pointed out the fundamental limitation of any cell analysis.

Practical Cell Analysis   Dimitri Pappas
© 2010 John Wiley & Sons, Ltd

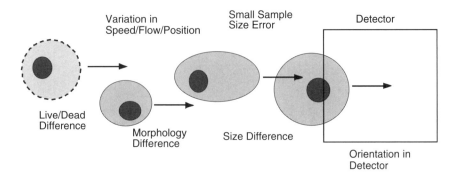

Figure 8.1 Sources of variation in an example cell measurement. If cells pass in a laminar fluid stream through a detection volume (optical, electrical, etc.), each cell will produce a characteristic signal used to generate a mean value. The variation in the mean will be the cumulative effect of variations in cell flow, cell properties, orientation in the detector, instrumentation noise, and other sources of error. Sources of instrument and mechanical noise, and sampling error must be minimized to elucidate an accurate mean and the chemical/biological variation of the cells

While Gosset's work focused on the noble art of brewing at the Guinness Brewery, his observations were of such importance that they are taught today as introductory canon. However, it is often surprising to see a well-executed manuscript for peer review that has either ignored or mistreated sampling and statistics. The goal of this chapter is to refresh the experienced researcher or provide a primer for students on the methods of statistical reporting of cell-analysis data. Examples related to previous chapters are used when applicable.

8.2   TYPES OF ERROR

Consider the diagram of cells entering a detection zone in Figure 8.1. The detection zone could be a Coulter orifice (Chapter 6), a focused laser beam (Chapters 4 and 6), or camera field of view. If the cells are flowing toward the detector, and each cell response is measured sequentially, a mean (average) value of the cell measurement can be made, based on the sum of each cell response:

$$\bar{x} = \frac{\sum R}{n} \approx \mu. \tag{8.1}$$

In Equation 8.1, R is the response of each cell (size, fluorescence, etc.) and n is the number of cells measured. When measuring the *entire population*, μ can be used to represent the mean. When a *representative sample* of the population is measured, the sample mean $\bar{x}$ is used. In practice, the entire population is rarely sampled, so $\bar{x}$ must be related to the mean mathematically (Section 8.3). The population mean μ can be used when the population is small and every member of the population has been analyzed. For example, when discussing an exam average in a course, it is safe to use μ, as the entire class will have taken the exam. However, if that exam result is used to make an inference about students in general, then the sample mean should be used.

When producing a mean value from Gaussian events, there will be a standard deviation associated with it. The standard deviation is a reflection of the *error* or *uncertainty* of the measurement. The standard deviation is calculated based on the mean value and the difference between the mean and each sample value (x_i).

$$s = \sqrt{\frac{\sum (\bar{x}-x_i)^2}{n-1}} \approx \sigma. \tag{8.2}$$

Again, s is the standard deviation of the *sample* and σ is the standard deviation of the *population*. The same rules apply for determining if s or σ should be used. The standard deviation arises from the processes outlined in Figure 8.1, as well as any other variation or error in the sample. Using Figure 8.1 as an example, there are several sources of variation between cells entering a detection region. If the flow is laminar, then cells will have different velocities perpendicular to the fluid axis. If the dwell time of the cell in that detection zone affects the signal response, then laminar flow will produce an error. Differences in cell size may affect the measurement, as well as variation in cell morphology and cell orientation in the detection region. Cell viability may serve as a source of error in many measurements, as can differences in cell function, depending on the assay. Some of these variations, such as cell size, occur naturally and are important values in their own right. Others, such as flow effects, are introduced by the experiment design. These human-made errors obscure the chemically and biologically relevant variation and should be minimized. The final source of variation in Figure 8.1 is the detection region itself. What is the noise of the detector, and how does it affect signal? In any system, the noise adds quadratically and can be expressed on a global level as follows:

$$s_{\text{total}} = \sqrt{s_{\text{chem}}^2 + s_{\text{biol}}^2 + s_{\text{instrum}}^2 + s_{\text{sampling}}^2}. \tag{8.3}$$

In this case s_{chem} and s_{biol} are the random chemical and biological variation of the sample, respectively. The instrument (measurement) error $s_{instrum}$ is the quadratic sum of the random errors associated with the analytical process itself. The sampling error, $s_{sampling}$, is affected by sample size and nonrepresentative samples. In all experiments, these four major sources of variation are present at the same time. The chemical and biological variations are of interest in most experiments; therefore, the instrumentation and sampling error must be minimized *relative* to the chemical and biological relevance. For example, a flow cytometer with a 5% coefficient of variation (Section 8.3) can be used to examine differences in fluorescence intensity that have a biological variation on the order of 20–25% in some cases. For DNA content measurements, however, the same coefficient of variation (5%) is equal to or greater than the biological variation, and would obscure the true randomness of the sample.

Chemical and biological variation should not, in the strict sense, be considered noise. The deviation of values about a mean, when related to the randomness of nature, can be used along with the mean to describe the sample. Examples of chemical variation include convection or movement of analyte within a cell, both of which are indicative of the intracellular environment. Biological variations, such as differences in antigen density on a cell surface can be used to determine cell growth and response to stimuli. Instrument precision can be determined both intra- and inter-measurement by measuring identical samples containing identical analytes. For flow cytometry, microfluidics, and microscopy, calibrated beads with nearly identical fluorescence intensities can be used to calibrate size and fluorescence measurements. Other standards for these methods are listed in Chapter 9. If standards of uniform (monodisperse) size and intensity cannot be obtained, the inter-measurement precision can be estimated by performing repeat measurements of a given sample.

In many cases, the instrument error cannot be improved, particularly for "closed-box" instruments that have few user inputs or adjustments. Even when an instrument's precision can be adjusted (e.g., through laser alignment on a flow cytometer, or a cooled camera on a microscope), there will be a minimum error that cannot be improved. The only other source of variation that can be affected is the sampling error. As pointed out in the introduction, the work of Gosset [1] and others have outlined the dependence of sample size and the error associated with counting events. Whether counting cells, stadium attendees, molecules, or any other individual events, the error associated with counting these n events is

$$s = \sqrt{n}. \tag{8.4}$$

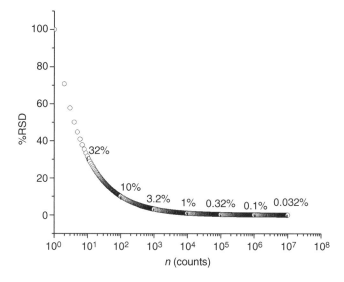

Figure 8.2 Error (presented as % relative standard deviation, %RSD) that arises from sampling *n* cells. The inverse square-root relationship dictates that increasing the sample size improves precision dramatically at first. However, small improvements are obtained after counts of several thousand cells

As the number of counted events increases, the error increases with a square root dependence. There are two important points to make about counting. First, increasing the number of counted events (e.g., the sample size) improves the relative error. Second, there is a point (Figure 8.2) where increasing the counts by a large number has a minimal effect on the total precision. At this point, the amount of additional time and effort spent to collect the additional cells may not be worth the smaller increase in precision. For example, the relative standard deviation $(s/\bar{x})$, expressed as a percentage, is equal to

$$\%\mathrm{RSD} = \frac{s}{\bar{x}} \times 100. \qquad (8.5)$$

Counting only 10 events (cells, votes, etc.) will yield an error of 32%. It is not difficult to extend this count range to include 100 events, lowering the error to 10%. However, increasing the count rate 10-fold to 1000 events will only result in an improvement of $10^{1/2}$, or 3.2%. Increasing to 10 000 events will produce a 1% counting error. It is at times tempting to increase the number of cell counts to the highest possible value, in order to get the best precision. However, a 1% error is typically better than the instrument precision of most analytical methods. Therefore, increasing the cell counts to the hundreds of thousands or millions is often unnecessary. It is important to note that, at this point, only counting error is being discussed. If a rare cell is to be detected, say a cell that is one in every 10^5 background cells, then higher *total cell counts* are required in order to obtain a statistically valid number of *rare cells*.

Figure 8.2 also outlines the fundamental drawback of microscopic measurements of cells. For all of the spatial resolution and information content of microscopy, it is inherently a few-cell technique, with many exposures of different fields of view required to obtain an adequate number of cells. Flow cytometry was developed to specifically combat this limitation, with cell counts in the tens of thousands occurring in a matter of a few minutes or less. Other techniques are, by virtue of the data collection process, slow and require many sample runs and a great deal of time to generate a large enough sample pool.

Tip: When considering how many cells are "enough," consider the other errors and variations in the sample, as well as experiment cost and time.

When sampling a population, the counting error reflects only the nature of counting random events. Any error produced by *nonrepresentative sampling* will be an additional source of variation in addition to the counting error. For example, when conducting research with primary cells, selecting only one portion of the tissue may affect experiment outcome, or only using cells from one animal. An example that is unrelated to cells, but serves to illustrate the issue of sampling, is the grading of exams. In many exam situations, students may turn in the exam when finished (rather than waiting until the exam period is finished). The pile of exams therefore bears a chronological history of the student completion order. Exams near the top (those that took the full exam time) will differ from those in the middle, and will be even more different from the earliest exams on the bottom of the stack. Even in this case there will be variation in the earliest exams, which will come from students who will receive full marks or who gave up in frustration with half of the exam incomplete. By choosing several random samples (and perhaps homogenizing the sample by shuffling exams), then a few exams could be graded and used to predict the outcome of the population. Careful cell sampling will reduce the sampling error to the limit of the counting error.

Instrument noise should, in the best cases, be the smallest source of noise, and nearly on par with the sampling noise. These conditions will allow the sources of variation mentioned so far to be elucidated. Instrument noise in many cell analyses will arise from light-source noise, detector noise, and mechanical (e.g., vibration) noise. In flow-based systems, variation (noise) in flowing cells will also add to the total uncertainty. Light-source noise (both shot and flicker noise) can be minimized via referencing the light source to monitor for intensity changes. Detectors, primarily photomultiplier tubes and CCD cameras, can be cooled to reduce dark noise and improve performance. Mechanical noise will affect light-source alignment and detection optics, and plays a large role in microscope-based measurements (particularly if microfluidic devices are used).

Instrument noise can be classified in several broad groups. Johnson (thermal) noise affects electronics and can be reduced by lowering the resistance of the internal circuitry or cooling. In both cases, there is a square-root improvement, so increasingly large changes must be made for modest improvements in the overall signal-to-noise ratio. In most analyses, the researcher does not have access to or control over the internal circuitry. If, however, cooling is an option, it should be used. Shot noise occurs from the random nature of photon emission and detection, as well as other random processes. Shot noise is a fundamental limit of the minimum instrumentation noise; it cannot be eliminated. Like counting error, the noise from shot effects is equal to the square root of the signal. Therefore the effect of shot noise can be minimized if one considers the signal-to-noise ratio:

$$\frac{S}{N} = \frac{\bar{x}}{s} = \frac{\bar{x}}{\sqrt{\bar{x}}}. \tag{8.6}$$

It is important at this point to note that when the measurement is strictly limited by the instrument noise, the relative standard deviation is the inverse of the signal-to-noise ratio. Figure 8.2 can therefore describe the improvement in the signal-to-noise ratio if the number of counts, n, is substituted with the average signal $\bar{x}$. The signal-to-noise ratio, when limited by shot noise, can be improved by either increasing the signal or by *averaging more measurments*. As discussed in Chapter 4, it is not always practical to average more. Acquiring more averaged data requires more time. In the case of microscopy, it also results in increased photobleaching. When averaging is possible, it is best to determine how many averages yield a dramatic change in the signal-to-noise ratio, and not acquire beyond that point.

Flicker noise differs from shot noise because it is both frequency dependent and linearly proportional to the signal. If the signal increases 10-fold (or 10 averages are taken), then the signal-to-noise for shot noise increases by $10^{1/2}$. However, the flicker noise will increase 10-fold and *there will be no improvement in the signal-to-noise ratio*. To determine if flicker noise is the dominant source of noise, the noise can be plotted as a function of increasing signal (if possible) or increasing numbers of averages. If the noise increases linearly, then there is no benefit from averaging or improving the signal. Instead, the source of flicker noise should be determined. Flicker noise occurs from sources such as mechanical vibration, long-term power-supply drift, and convection of the sample, among others. Referencing signals can minimize some flicker noise.

Line or interference noise arises from the alternating current from power lines. In the United States, power lines operate at 60 Hz; in the European Union 50 Hz power is more common. Therefore, signals obtained at frequencies that are multiples of the power line frequency (e.g., 60, 120, 180, 240, etc., in the United States) will suffer from interference noise. Measurements taken at other frequencies will not pick up this interference and should be used instead.

8.3   FIGURES OF MERIT IN STATISTICAL ANALYSIS OF CELLS

There are many figures of merit that can be used to discuss statistical values, including the mean and standard deviation already discussed. The relative standard deviation, RSD, is also expressed as the coefficient of variation (CV or %CV)

$$CV = \frac{s}{\bar{x}} \approx \frac{\sigma}{\mu}. \tag{8.7}$$

The %CV is simply the CV multiplied by 100%. The CV and RSD are valid for nonzero means; if the mean is negative, it should be stated with the CV or RSD to avoid confusion. The CV and RSD have advantages primarily when comparing techniques or methods, as they are dimensionless. When the CV is limited by instrumentation variability, then it is the inverse of the signal-to-noise ratio.

When discussing a sample containing many values, it is possible to use the mean or median. Each has its own advantages. The mean reflects the pooling of all values, including outliers. However, a handful of high or low values can skew the mean. This is particularly true of non-Gaussian (or non-normal) distributions, such as the Poissonian distribution (Figure 8.3). In the case of the Gaussian distribution, the median and the mean are identical values. If the Gaussian peak exhibits a tailing effect,

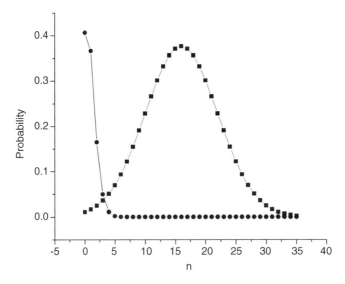

Figure 8.3 Representative Poissonian (circles) and Gaussian (squares) distributions. Gaussian events are random, while Poissonian events are random and rare (e.g., single rare cells or single molecules)

then the tail values skew the mean. In Poisson statistics, rare, random events such as single cells or single molecules will produce a distribution, which consists predominantly of small values (Figure 8.3). The Poisson distribution for rare events (such as cells) can be expressed as

$$P(k) = \frac{\lambda^{-k}e^{-\lambda}}{k!}, \tag{8.8}$$

where λ is the mean number of cells (or cells per unit volume) and k is the number of cells detected in the volume at any given time. If a concentration of one cell is in the detection volume, then 36% of the time there will be zero cells detected, and an equal percentage of the time one cell will be detected. Two cells will be in the same volume 18% of the measurement time. If the cell concentration is lower, such that only 0.1 cells are in the detection volume, then the probability of detecting zero cells is 90%. The probability of detecting one or two cells is 9% and 0.4%, respectively. In this case of 0.1 cells, the mean probability value is 2%; the median of the same data set is zero. The mean does not change in this case for reasonable values of λ, while the median reflects the distribution of data.

In flow cytometry, the median fluorescence intensity can be used for gated or ungated data to reject outlier values, such as brightly stained

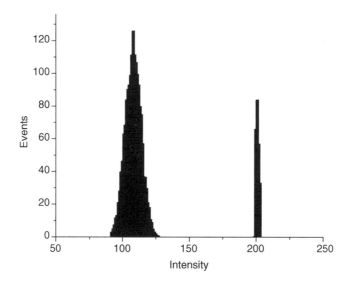

Figure 8.4 A conceptual histogram of cell intensity. The presence of some brightly stained cells near 200 counts (e.g., cell double occupancies, brightly stained debris) can skew the population mean, while the median is not affected to the same extent. If gated, the mean cell fluorescence would be 112, while the ungated mean is 121. If the median value is used, the fluorescence (gated or ungated) median is 112

debris. Figure 8.4 depicts simulated data with a cluster of pseudo-Gaussian data and several outliers. The mean value if the outlier data was absent would be 112. The median value with outliers present is 112, while the mean in this case is 121. This is, of course, a simulated case but it illustrates the point that the median is affected by outlier data to a smaller degree.

The mean of a sample $\bar{x}$ can be used to predict the mean of the population using confidence intervals and the Student's t value

$$\mu = \bar{x} \pm \frac{ts}{\sqrt{n}}. \qquad (8.9)$$

The t value (Table 8.1) is chosen for both the degrees of freedom $(n-1)$ and the confidence level. When reporting a cell population mean using this method, both the number of cells or trials (n) and the confidence level should be reported.

Tip: As in any case where a number of cells or repeat trials are used, there is a square-root return on time and effort spent improving the confidence interval.

Table 8.1 Student's *t* values for confidence intervals and t-test (note, only 90, 95, 99, and 99.9% confidence values are shown)

Degrees of Freedom ($n - 1$)	Confidence (%)			
	90	95	99	99.9
1	6.314	12.706	63.657	636.619
2	2.920	4.303	9.925	31.598
3	2.353	3.182	5.841	12.924
4	2.132	2.776	4.604	8.610
5	2.015	2.571	4.032	6.869
6	1.943	2.447	3.707	5.959
7	1.895	2.365	3.500	5.408
8	1.860	2.306	3.355	5.041
9	1.833	2.262	3.250	4.781
10	1.812	2.228	3.169	4.587
20	1.725	2.086	2.845	3.850
30	1.697	2.042	2.750	3.646
40	1.684	2.021	2.704	3.551
50	1.676	2.009	2.678	3.496
100	1.660	1.984	2.626	3.390
120	1.658	1.980	2.617	3.373
∞	1.645	1.960	2.576	3.291

When plotting data, the error of the value should always be presented as bracket error bars. However, which figure of merit best expresses the variation of the mean value? The answer, in short, is that there is some flexibility in choosing what error is used. Since different figures of merit can be used to assign error bars, one should always state what type of error is used in the manuscript, figure, and so on. There are several options when plotting error bars or stating error around a mean in tabular form. The standard deviation, relative standard deviation, confidence interval, and standard error of the mean can be used. The standard deviation and RSD have already been defined and serve to express the variation in the sample values around the sample mean. It is best not to use the RSD for error bars, as the mean will have units, but the RSD will be dimensionless. If standard deviation is used, the number of deviations should also be reported. For example, one may express an error as multiples of the standard deviation in some cases.

When expressing the variation of data due to chemical or biological variation, the standard deviation is a good reflection of the data around a mean. The confidence interval should be used when the goal of the analysis is to present a mean value that is as close as possible to the population mean, and where the chemical and biological variation is not as important. For example, when measuring cell size by microscopy, the standard deviation is an important data point in its own right. For analysis of a particular cell count for an individual patient by flow cytometry, the population mean (i.e., the patient's blood-cell count) is more important. Confidence intervals, calculated using the Student's t value that matches the degrees of freedom and confidence level, can be used to express the accuracy of the true mean. While different confidence levels are used, a confidence of 95% is a good compromise between certainty and unreasonably large intervals. Confidence intervals can be used especially when comparing cell-analysis methods against a standard sample or standard method.

The standard error of the mean can also be used to express error. The standard error is

$$\mathrm{SE}_{\bar{X}} = \frac{s}{\sqrt{n}} \approx \frac{\sigma}{\sqrt{n_{\mathrm{pop}}}}, \qquad (8.10)$$

where s and σ are the sample and population means, respectively; n is the sample size and n_{pop} is the size of the entire population, if it can be measured. The standard error of the mean corrects for sample size effects on the standard deviation. When counting error is the only source of uncertainty, the standard error of the mean is unity, or 100%. Therefore the standard error of the mean is useful when the source of error does not arise from counting statistics. Like the confidence interval, the standard error of the mean is more useful when discussing the accuracy of the sample mean, rather than the chemical and biological variation of the cells.

8.4   LIMITS OF DETECTION AND QUANTITATION (OF CELLS)

In Chapter 5 the concept of a limit of detection for cells was introduced. In chemical analysis, the signal-to-noise is defined by the International Union of Pure and Applied Chemistry (IUPAC) as

$$N_{\mathrm{L}} = 3s_{\mathrm{B}} = 3\sqrt{n_{\mathrm{B}}}. \qquad (8.11)$$

In the case of cell analysis, the standard deviation of the blank (s_{B}) can be replaced by the number of cell counts in a control sample (n_{B}). Table 8.2 lists practical approaches to measuring s_{B} in different cell methods, such as flow cytometry and affinity separations.

Table 8.2 Methods for measuring the background cell noise (s_B) by analytical method

Method (chapter)	Blank or control cell measurement
FACS (5)	Sort nontarget cells using antibody labels. Sorted cell count (n_B) used to calculated s_B.
Flow Cytometry (6)	Label nontarget cells and use median fluorescence to set gate. Measure events in positive gate (where target cells would be) to obtain n_B.
Affinity Separations (5)	1. Measure target cells in nonaffinity separation (i.e., binding chemistry without capture molecule) to obtain nonspecific binding and n_B. 2. Measure target cells with mismatched capture molecule to obtain n_B. 3. Measure nontarget cell with capture molecule for target cell to obtain n_B (preferred).
Fluorescence Microscopy (4)	Measure nontarget cells and obtain cell count of positively stained cells (n_B, i.e., incorrectly identified nontarget cells).

The background cell count varies by the type of method. For example, in flow sorting, a group of target cells is being sorted out of a presumably larger population of non-target cells. The number of positively sorted cells must be larger than the background (non-target) cells by a statistically relevant number. The limit of detection is therefore the minimum number of cells that can be measured as larger than the non-target cells randomly captured. In order to quantify cells (i.e., report a concentration), the limit of quantitation should be used [2]. The limit of quantitation is

$$N_Q = 10s_B = 10\sqrt{n_B}. \tag{8.12}$$

The limit of quantitiation is higher than the limit of detection, since more cells are needed to report a concentration than are required to report the *presence* of a particular cell type. These limits are the *minimum* cell counts needed to detect or quantify a target cell, respectively. Whenever possible, these limits should be exceeded as much as possible to improve measurement statistics. The limits are, however, useful in comparing analytical methods, and also in gauging the overall potential of a technique. For example, consider a cell separation method that exhibits 0.3% nonspecific binding. This level of nonspecific binding will impose limits on how rare the target cell can be before it cannot be detected reliably. If erythrocytes are removed by lysis (Chapter 5), then the rare cell will be present among the leukocytes. If an example concentration of 10^6 cells ml^{-1} is used, then a nonspecific binding of 0.3% will result in 3000 cells nonspecifically captured per ml of sample. In most cases, smaller volumes of blood are typically analyzed, so in a 20 µl sample 60 leukocytes will be captured by nonspecific binding by the separation method. The "noise" or deviation in this nonspecific cell capture is the square root of the background cell count, or eight cells. The rare cell must be present at concentrations of 24 cells to be *detected* at all. In this minimum case, a total of 84 cells would be captured, with 24 being target cells and 60 non-target cells. The 24 cells in a 20 µl sample would yield a concentration of 1200 cells ml, or 0.12% of the original blood sample. In order to quantify the rare cells, the more stringent limit of quantiation must be used, and the minimum rare-cell concentration that can be quantified is 4000 cells ml^{-1} or 0.4% of the total sample. If a rare cell is found on the order of one target cell for 10^5–10^7 background cells, then most methods cannot reliably detect the presence of the rare cells, let alone perform additional analyses upon them.

The limits of detection and quantitation are useful to set constraints on how efficient the separation (or other analysis) is at measuring target cells and rejecting nontarget cells. From a practical standpoint, these simple calculations can save weeks of failed research attempting to measure or isolate rare cells. From the standpoint of a proposal or manuscript reviewer, these same calculations can be used to assess the feasibility of a proposed idea.

8.5   METHODS TO IMPROVE CELL STATISTICS

There are steps that can be taken to improve the statistics of a cell analysis experiment. One must, however, consider what the sources of noise/error are and whether they can be minimized. Taking Figure 8.1 into account, there are variations from the sample and the measurement process. In all cases, the instrumentation noise should be minimized. When the chemical and biological variation is important, a small instrument error will allow these natural variations in the sample to be elucidated. Of course, if sampling error is present in this case, the natural variation will be convoluted with the sampling error. If the goal of the experiment is to present a sample mean that is accurate with respect to the population mean, then instrumentation and sampling error must be small, and the sample size error should also be negligible.

Instrument error can be minimized via upgrading to higher-end, cooled detectors, or referencing laser or mechanical noise. This is not always possible, especially if the instrument cannot be modified. If the instrument error is limited by shot noise, then averaging can be used to improve precision. As mentioned in Chapter 4, averaging in some cases can degrade the sample, such as in repeated bleaching in microscopy. Averaging, however, only improves some sources of error. Errors in sampling can be improved by increasing the sample size.

Consider, for example, measurements of cells stained and analyzed by fluorescence microscopy. In the simplest case, these cells may be considered brightly or dimly stained, depending on their intensity and a determined threshold. The threshold could be a multiple of standard deviations above the background mean (e.g., the limit of quantitation) or determined by other means. The sampling error is then governed by counting statistics, assuming the fluorescence intensities are not widely scattered and the instrument precision is high. Using a $10\times$ objective, only several dozen cells can be counted in a particular field of view. Acquiring several images, each with a different field of view, will increase the total sample number and improve the counting statistics at the expense of analysis time. Switching to a higher speed or high cell-count analysis, such as flow cytometry (Chapter 6) or certain cell separations (Chapter 5) will improve counting statistics and in many cases reduce analysis time. However, the urge to choose the fastest analysis or the one that can produce the largest cell count must be tempered with the desired goal. If, for example, morphology is important, then higher magnification (and lower cell count) microscopy will be required, and the sampling statistics will be limited by the number of analyses made and time available.

8.6   COMPARING ANALYTICAL VALUES

When reporting two values and their respective differences, the statistical significance of those differences must also be presented. Whether comparing two methods, disputing results, or evaluating the accuracy of a cell measurement relative to standards, the differences in the mean values are not valid unless proven so. In most cases, two values, each with their own uncertainty, will be compared. To do this, a t-test is performed, using the Student's t values. The t-test compares a calculated t value against the tabulated t value for 95% confidence and $n_1 + n_2 - 2$ degrees of freedom (here n_1 and n_2 are the sample numbers, respectively, of the two means).

The 95% confidence value is chosen so that the claim that two values are equivalent is only wrong once in 20 times, an acceptable level of error. The calculated t value is given by

$$t_{\text{calc}} = \frac{|\bar{x}_1 - \bar{x}_2|}{s_{\text{pooled}}} \sqrt{\frac{n_1 n_2}{n_1 + n_2}} \qquad (8.13)$$

where $\bar{x}_1$ and $\bar{x}_2$ are the two values being compared, n_1 and n_2 are the sample numbers of the two means, and s_{pooled} is the pooled standard deviation defined as

$$s_{\text{pooled}} = \sqrt{\frac{s_1^2(n_1 - 1) + s_2^2(n_2 - 1)}{n_1 + n_2 - 2}}. \qquad (8.14)$$

The standard deviations s_1 and s_2 correspond to $\bar{x}_1$ and $\bar{x}_2$, respectively. The calculated t-value must be larger than the tabulated value in order to state that the two mean values are statistically different. If two values are significantly different at higher confidence levels, this should be stated.

Example: The mean fluorescence intensity of stained cells in four separate flow cytometer measurements is 28 with a standard deviation of 17 (arbitrary units). The control cells (four measurements) have a mean fluorescence intensity of 12 with a standard deviation of 8 (arbitrary units). Is the difference in fluorescence intensity between the stained samples and controls statistically significant?

The pooled standard deviation is

$$s_{\text{pooled}} = \sqrt{\frac{s_1^2(n_1 - 1) + s_2^2(n_2 - 1)}{n_1 + n_2 - 2}} = \sqrt{\frac{17^2(4 - 1) + 8^2(4 - 1)}{4 + 4 - 2}} = 13 \qquad (8.15)$$

and the calculated t-value is

$$t_{\text{calc}} = \frac{|28 - 12|}{13} \sqrt{\frac{4 \times 4}{4 + 4}} = 1.74. \qquad (8.16)$$

The tabulated t-value for six degrees of freedom is 2.447, so the two results are *not* statistically significant.

Table 8.3 Q test values for rejecting data [3]

Q (90%)	n
5	0.64
6	0.56
7	0.51
8	0.47
9	0.44
10	0.41
15	0.34
20	0.20

8.7   REJECTING DATA: PROCEED WITH CAUTION

Outlier values in a sample population can drastically affect the mean, and can be removed if they are subjected to a Q test. The Q test compares the difference between the outlier and its nearest neighbor, and the entire range of data. Like the t-test, a calculated Q value is compared against a tabulated one (Table 8.3, given at the 90% confidence level). The calculated Q value is given by

$$Q_{\text{calc}} = \frac{\text{Gap}}{\text{Range}} = \frac{\text{Outlier} - \text{Neighbor}}{\text{Largest} - \text{Smallest}}. \tag{8.17}$$

The outlier must, by definition, be either the smallest or largest value. If data is presented that was subject to the Q test to remove an outlier, it should be stated as such. The Q test should only be performed once (indeed, it becomes increasing difficult to reject additional data points).

The Q test should only be used when there is a clear outlier, with a valid suspicion of why a systematic error has occurred. One must question how often the Q test is being applied in an experiment. The Q test should be applied sparingly. In fact, data that must be routinely subjected to the Q test is likely to be influenced by a systematic error that should be discovered.

8.8   CONCLUSION

Cell measurements require an understanding of the many sources of variation in a measured value. The natural variations of cell size, cell type, antigen density, cell cycle, viability, and so on, are in their own right valid parameters to consider. Noise from the instrument, sampling error, and other sources can obscure the natural variations and also reduce the accuracy of the mean. Methods to improve the precision of cell measurements, such as higher cell-count methods and averaging (when applicable) can help to elucidate the true nature of the sample. The accurate reporting of statistical data aids in the proper dissemination and evaluation of any scientific work. The correct use of tests to report differences or reject data provides meaningful analysis that can enhance the impact of cell analyses and results.

REFERENCES

1. Student. (1907) On the error of counting with a haemacytometer. *Biometrika*, 5, 351–360.
2. Long, J.L. and Winefordner, J.D. (1983) Limit of detection, a closer look at the IUPAC definition. *Analytical Chemistry*, 55, 712A–724.
3. Harris, D.C. (2005) *Exploring Quantitative Analysis*, 3rd edn, W.H. Freeman, New York.

9
Protocols, Probes, and Standards

9.1 INTRODUCTION

The purpose of this chapter is to compile many of the protocols discussed in this book into one section for easy reference. Some protocols were placed in their respective chapters for emphasis or ease of use. The protocols, probes, and standards listed in this chapter correspond to techniques discussed in previous sections, and are grouped according to chapter topic. The protocols compiled herein are not an exhaustive list of every technique, but rather contain many common analyses and can serve as a guide for modification or further investigation. Mention of specific manufacturers or vendors does not convey any endorsement or the implication that other, suitable substitutions will not work as well. As in any protocol, reagent concentrations, reaction/incubation times, and so on, should be adjusted for the experiment at hand.

9.2 CELL TRANSFECTION AND IMMORTALIZATION (CHAPTER 1)

There are several approaches to transfecting mammalian cells with plasmid DNA. Viral vectors can be used, although the required biosafety level may exceed the BSL rating of the laboratory. The use of plasmid

Practical Cell Analysis   Dimitri Pappas
© 2010 John Wiley & Sons, Ltd

delivery agents and methods, such as polyamine delivery or electroporation, are easier to implement in most laboratories. When performing a
transfection, it is important to not only optimize the protocol for the best
(high yield) transfection, but to also provide an assay that will assess the
degree of transfection.

PROTOCOL 9.1: TRANSFECTING CELLS WITH POLYAMINE REAGENTS

This protocol uses a polyamine delivery agent such as cadaverine. Polyamine reagents are commercially available, such as the GeneJammer
product from Stratagene. The exact polyamine used in the GeneJammer
protocol is proprietary, but the general mechanism of delivery takes
advantage of the polyamine transport system present in many mammalian
cells. Since the phosphate backbone of DNA has a high density of anion
groups, the highly charged, cationic polyamine will therefore transport
the DNA into the cell.

Cells: Adherent or suspended cells can be transfected by this method.
Cells should be in the log phase of growth, with high viability. Cells can be
grown in complete medium, and can be transfected in serum or serum-free
medium. The transfection should be attempted in serum-free medium
first. This method has been used to transfect a variety of cells with
reporters, including FRET-based reporters for caspase activity in
apoptosis [1].

Reagents needed: Plasmid DNA (highly purified), polyamine delivery
agent, serum-free medium. Note: Mix all sterile reagents in a biosafety
cabinet.

1. Three control groups are needed, plus any additional samples for
   concentration optimization. A blank control should be handled
   (washed, incubated, etc.) using the same methods, but without the
   polyamine reagent or DNA construct. The polyamine control
   should likewise be handled, but should contain the polyamine
   reagent, and not the DNA construct. The final, DNA control should
   be treated with the DNA construct only and not the polyamine
   reagent. All other samples should contain both DNA and reagent.
2. The ratio of polyamine reagent and DNA should be optimized. For
   the GeneJammer system, a ratio of polyamine reagent to DNA
   should be $3:2-6:1$. For other products, this concentration may

vary. In any case, several concentrations in this range should be tested to obtain the highest transfection efficiency.

3. Adherent cells should be grown up to 80% confluent in the culture dish; adherent cells should be at a concentration of 10^4–10^6 cells ml^{-1} (see Chapter 2 for cell concentration measurements using a hemacytometer).

4. Basic medium (serum-free) should be mixed with the polyamine reagent in a sterile plastic tube at a 3% (v/v) concentration of reagent. Note: Polyamine reagent will adhere to the plastic of the tube and should be added directly into the medium. Incubate 5–10 minutes at room temperature.

5. Add DNA at the appropriate ratio (3 : 2–6 : 1) and incubate 15–45 minutes at room temperature.

6. Place cells in a biosafety cabinet. Note: it is important that only one cell line be worked with at a time to avoid cross-contamination.

7. Add the medium containing the delivery reagent and DNA to the cell cultures. The concentration (v/v) of delivery medium over the total volume will be 2.5–6.0%. Mix by gently rocking before returning the culture to an incubator. Note: Ensure sterility at all times of the transfection process.

8. If the cells are maintained in serum-free medium, replace with complete medium after 3–8 hours.

9. Culture cells for 24–72 hours and assay for the desired phenotype. If fluorescent protein reporters are part of the plasmid construct, then flow cytometry (Chapter 6) or fluorescence microscopy (Chapter 4) can be used to assay the relative expression of the plasmid.

10. Cells can be selected using fluorescence-activated cell sorting (FACS) or other cell-separation methods (Chapter 5) to isolate a phenotypically pure cell line expressing the new proteins.

Note: This method will produce transient transfection. That is, the additional plasmid DNA is lost after several growth cycles. In order to produce a permanent addition to the cell genome, see Protocol 9.2 for Stable Transfection using Polyamine Delivery.

PROTOCOL 9.2: STABLE TRANSFECTION USING POLYAMINE DELIVERY

Stable transfection requires that the cell line incorporates the plasmid into its genome, rather than lose the construct over several cell cycles

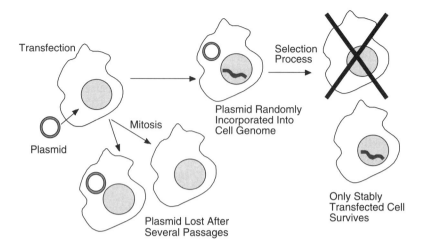

Figure 9.1 Transfection of cells using plasmid DNA typically results in transient expression of the gene until the plasmid is lost via mitosis. In stable transfection, the few, rare cells that incorporate the plasmid into the cellular genome are selected by growing cells in culture medium containing a selection agent

(Figure 9.1). In general transfection, a DNA construct is added and the desired phenotype is monitored. Cells expressing the required phenotype are then used for the experiment, but they may not have permanently incorporated the genes. To stably transfect the cell, the plasmid must contain a second gene that introduces resistance to a particular selection antibiotic. Positively transformed cells will resist the antibiotic while those that are not transfected die. The cells are then cultured for 1–2 weeks. This long culture time ensures that transiently transfected cells will also die, and that only stable cells will survive.

1. Neomycin analog G418, also called Geneticin, is commonly used as a selection agent [2]. The bacterial gene *neo* imparts Geneticin resistance to mammalian cells. Other selection reagents can also be used, provided a gene can be inserted that imparts some form of resistance or survival to the transformed cells.
2. Determine the required antibiotic concentration to kill nontrans-fected cells in culture by growing 5–10 cell samples in microwell plates or dishes and titrating the concentration of the selection agent. Check daily for viability and replace the selection medium (complete medium with selection agent present) daily to remove cell debris.
3. Transfect cells according to Protocols 9.1 or 9.3 (or other transfec-tion methods).

4. Culture transformed cell samples in complete medium without selection agent for 1–2 days.
5. Passage cells for 1–2 weeks in selection medium. Replace medium daily.
6. Monitor viability during the passages with viability probes (Table 9.2) and visual inspection on a fluorescence microscope (Chapter 4).
7. After 1–2 weeks, only cells that have been stably transfected will remain viable in the selection medium. Perform an assay (Protocol 9.1) to determine that the desired gene has been introduced into the cell line. Cell separation will not be needed, as all cells should now express both the desired gene and the resistance gene.

    Note: For slow-growing cells, longer culture times may be needed (e.g., 2–3 weeks).
8. Culture cells in complete medium without selection reagent. Cells should be cryopreserved and the desired phenotype assayed periodically as a particular culture is passaged. If the phenotype changes, check for resistance to the chosen antibiotic. If resistance is lost, or the desired phenotype is no longer present, use a cryopreserved stock.

PROTOCOL 9.3: TRANSFECTION USING ELECTROPORATION

Electroporation was introduced in Chapter 1 as a method to introduce plasmid DNA into cells. A pulsed electric field produces temporary holes in the cell membrane, allowing foreign DNA to enter. The benefit of electroporation is that cell lines that are not amenable to other reagent-based DNA delivery methods can be transfected. In addition, electroporation can be a reagent-free method, where only the plasmid DNA is needed. One drawback of electroporation is that other material present in the transfection buffer or medium can also enter the cells. Another potential problem with electroporation is that the technique must be optimized for each cell line to ensure maximum viability and transfection efficiency. This protocol is adapted from the work of Baum and co-workers [3], and can be used for a variety of mammalian cell lines. As with all transfection experiments, sterility must be maintained at all times.

Reagents needed: Plasmid DNA containing the target gene (plus the gene for antibiotic resistance, if stable transfection is desired).

Cell Suspension   Electrodes

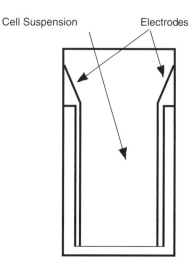

Figure 9.2   Electroporation cuvettes. Two planar electrodes define the electric field for electroporation, and are available with different gap spacing between the electrodes

1. For suspended cells, centrifuge for 5 minutes and resuspend in complete medium. For adherent cells, remove from flask or dish using trypsin (Chapter 3) before the centrifugation step. Cell density should range from 10^5–10^6 cells ml^{-1}.
2. Transfer cells to electroporation cuvettes with a 4 mm gap (Figure 9.2).
3. Add plasmid DNA to electroporation cuvettes. A control cell sample should not contain plasmid DNA. DNA should be on the order of 10 μg in 40 μl of buffer.
4. If the optimal electroporation protocols are known, skip to step 6.
5. The electroporation device must be optimized for both field strength and time constant. The optimization must balance cell transfection efficiency as well as cell viability.

   When optimizing cell transfection efficiency, a voltage of 200–300 V should be tested in 10 or 20-volt increments. To test for transfection efficiency, a membrane impermeant molecule may be used before experiments with plasmid DNA. A second dye for cell viability should be added after cells have been removed from the electroporation field.

   The sub-protocol for loading/survival optimization is given as follows:

Table 9.1   Impermeant dyes and probes for transfection analysis

Dye	Notes
Lucifer Yellow	Used extensively as a probe for delivery into cell
Fluorescein-labeled dextrans	Various molecular weights available. Other dyes may be used to label the dextrans
Calcein (not Calcein-AM)	The nonacetoxymethyl ester form of calcein is highly charged and does not enter the cell. It can be purchased in its free form or hydrolyzed from calcein-AM using KOH
Sulforhodamine 101	Less interference from cell autofluorescence and emission from green fluorescent protein

Note: Dead cells with compromised membranes will not exclude these dyes. A wash step is recommended, unless an impermeant viability dye (Table 9.2) is added after the transfection event and before analysis of transfection efficiency.

Optimizing Electroporation Parameters

a. Place cells as in steps 1–3 in an electroporation cuvette. Add a membrane-impermeant dye (Table 9.1) at a final concentration of 1–5 μM. This dye is the electroporation probe.

b. Vary the electroporation voltage from 200–300 V, with each sample increasing by 10–20 V.

c. Remove the test sample, add 1–5 mg ml^{-1} of propidium iodide or 7-AAD (Table 9.2) and stain in the dark for 5 minutes at room temperature. This dye is the viability probe.

d. Analyze cells by fluorescence microscopy or flow cytometry. Dead cells will stain positive for the viability probe, and may or may not stain positive for the electroporation probe. Live cells that were not electroporated will not stain for either the electroporation or the viability probe. Live, electroporated cells will stain positive for the electroporation probe and will not stain for the viability probe. In most cases, a balance between the brightest fluorescence from the electroporation probe and the fewest dead cells must be reached. These electroporation parameters can then be used to insert DNA or other materials into cells.

6. Using the optimized electroporation parameters, pipette the cell sample into buffer containing plasmid DNA and electroporate the cells.

7. Remove cells from the cuvette by transferring the suspension via a pipette to a new culture vessel. Rinse the cuvette twice with complete medium and transfer rinsate to the same culture vessel.

8. Wait 12–24 hours before beginning the selection procedure if stable transfection is required.

Table 9.2   Probes for cell viability and/or DNA content

Dye	Application	Stain type	Compatible lasers (wavelength, nm)	Notes
Propidium Iodide	Fluorescence	Impermeant	Argon ion (488), Nd:YAG (532), HeNe (543)	Widely used for viability, works with 488 nm lasers
7-Aminoactinomycin D (7-AAD)	Fluorescence	Impermeant	Argon ion (488), Nd:YAG (532), HeNe (543)	Works well with phycoerythrin and 488 nm lasers
Trypan Blue	Colorimetry	Impermeant	N/A	Used in white-light microscopy
Calcein-AM	Fluorescence	Permeant, fluorogenic	Argon ion (488)	Dye cleaved into fluorescent product by viable cells. Other calcein variations available
Carboxyfluorescein diacetate	Fluorescence	Peremeant, fluorogenic	Argon ion (488)	Mechanism similar to Calcein-AM
SYTOX Green	Fluorescence	Impermeant	Argon ion (514), Nd:YAG (532)	
SYTOX Orange	Fluorescence	Impermeant	Nd:YAG (532), HeNe (543)	High quantum yield, also used as sensitive stain for DNA measurements
Ethidium Bromide	Fluorescence	Impermeant	Argon ion (514), Nd:YAG (532)	Can be combined with Hoechst dyes for viability measurements while excluding cell fragments and non-nucleated bodies
Hoechst 33342, 33258, 34580	Fluorescence	Permeant	Argon ion (351), Nd:YAG (355)	Permeant, A-T-selective DNA stains
DAPI	Fluorescence	Permeant	Argon ion (351), Nd:YAG (355)	Permeant, A-T-selective DNA stains

PROTOCOL 9.4: CELL IMMORTALIZATION USING hTERT TRANSFECTION

A Number of transfection methods can be used to insert the hTERT gene into a cell. Since the generation of an immortal cell line is the end result (i.e., not transient transfection), a resistance agent should be used to select stably transfected cells. Commercial sources such as Addgene (www. addgene.org) offer hTERT plasmids containing neomycin, puromycin, and hygromycin [4]. A plasmid for hTERT is also available from the American Type Culture Collection (www.atcc.org) as well as other vendors.

It is important to note that hTERT transfection does not immortalize all cell types. A longer analysis time is also required to ensure that the selected cells (if a selection resistance gene is included) do not become senescent.

1. Before transfection, culture a sample of the cells through as many passages as possible until the cells die or become senescent (see Figure 9.3). This maximum passage number must be determined and exceeded by positively transfected cells.

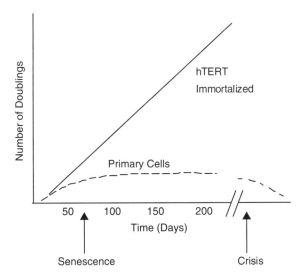

Figure 9.3 Conceptualized population doublings after initial cell isolation. In non-proliferating cells, the population reaches senescence (i.e., absence of cell cycle) after a short number of doublings. Crisis is reached when the senescent population suddenly dies off. In immortalized cells, the population continues to double indefinitely. (Adapted from [5])

2. Transfect the hTERT plasmid (plus selection gene) into the cell line using Protocols 9.1–9.3 (or other method of gene insertion). It is possible to monitor transfection efficiency using a probe from Table 9.1, although the incorporation of the probe may affect transfection efficiency.

3. Transfer transfected cells to the appropriate growth medium and culture for 12–24 hours.

4. Transfer cells to medium containing the selection agent. Culture for 1–3 weeks, changing the medium every 1–2 days.

5. Monitor the number of cells as a function of time (Figure 9.3). Stably transfected, immortalized cells should continue to double well beyond the point of senescence or crisis determined for nontransfected cells.

6. Freeze several stocks of the stably transformed cells as reference for the new cell line.

9.3 CALCULATING RELATIVE CENTRIFUGAL FORCE (RCF) AND CENTRIFUGE ROTOR SPEED (CHAPTER 2)

As discussed in Chapter 2, many centrifuges list the rotor speed in revolutions per minute (RPM). The centrifugal force, however, depends not only on the rotor speed, but also the radius. The relative centrifugal force, given as multiples of the gravitation force (g), can be calculated by the following:

$$g = 1.118 \times 10^{-5} rs^2 \tag{9.1}$$

where r is the rotor radius (in cm) and s is the speed in RPM. For fixed-angle rotors, the radius is determined by the distance from the center of the rotor to halfway through the sample tube (see Chapter 2, Figure 2.5). For swing-bucket rotors, the radius is measured as the distance to the sample.

9.4 FLUORESCENCE METHODS (CHAPTERS 4 AND 6)

Fluorescence methods, discussed in Chapters 4 and 6, provide sensitive biochemical analyses of intact cells and their contents. It is possible, for

Table 9.3   Caspase roles and probes

Amino-acid sequence	Target caspase[a]	Caspase role
VAD	All	n/a
DEVD	Caspase 3	Effector caspase
VEID	Caspase 6	Effector caspase
AEVD	Caspase 7	Effector caspase
IETD	Caspase 8	Initiator caspase; external (Fas) response
LEHD	Caspase 9	Initiator caspase; internal (cytotoxic) signal response

[a] The target caspase has the highest affinity for the complementary sequence, but other caspases can cleave the same sequence [8].

example, to measure intracellular ion concentrations (Table 9.4), intensity, size, antigen density (Table 9.5), and other parameters. For non-fluorescent analytes, conjugation with a fluorophore of choice is possible (Table 9.6). This approach allows a bright fluorophore with the desired spectroscopic properties to be conjugated to a protein or other molecule of interest. Some of the protocols listed in this section can be used with multiple instrumental techniques, while others are confined to a specific analytical method.

PROTOCOL 9.4: APOPTOSIS DETECTION USING FLUOROPHORE-CONJUGATED ANNEXIN-V AND A VIABILITY DYE

Apoptosis can be identified by several hallmarks based on biochemical or morphological changes during the cell-death process. In healthy cells, the phospholipid phosphatidyl serine is strictly located on the inner cell membrane. In mid- to late-stage apoptosis phosphatidyl serine externalizes and serves as a receptor for phagocytes. The calcium-binding protein Annexin-V has a high affinity for phosphatidyl serine when a moderate concentration of Ca^{2+} is also present [6]. Fluorophore-conjugated Annexin-V is available for a number of fluorescent labels, including fluorescein (and related dyes), phycoerythrin, Cy5, and a host of other dyes. Annexin-V will bind to externalized phosphatidyl serine, but will also bind to the inner phosphatidyl serine if the cell membrane is compromised. Dead cells – regardless of whether apoptosis or necrosis was responsible – will therefore positively stain for Annexin-V as well. Adding a membrane-impermeant DNA viability dye (Table 9.2) will allow dead and apoptotic cells to be distinguished (Figure 9.4). Annexin

Table 9.4 Common probes for ions in cells.[a]

Dye	Target ion	Stain type	Compatible lasers (wavelength, nm)	Notes
Fluorescein	H^+ (pH)	Permeant if produced by cleavage of fluorescein diacetate	Argon ion (458, 488)	Fluorescein excitation is affected by pH. Emission spectrum is unchanged
Carboxyfluorescein	H^+ (pH)	Permeant if produced by cleavage of carboxyfluorescein diacetate	Argon ion (458,488)	Carboxyfluorescein is better retained in cells than fluorescein after cleavage of acetate groups
SNARF-1-acetoxymonoester (AM)	H^+ (pH)	Permeant. Cleavage of AM group produces SNARF-1	Argon ion (488, 514), Nd:YAG (532), HeNe (543)	pK_a better suited to physiological range than fluorescein
Fura-2	Ca^{2+}	Permeant	N_2 (337), Nd:YAG (335)	Ratiometric
Indo-1	Ca^{2+}	Permeant	Diode (405)	Ratiometric
Fluo-4	Ca^{2+}	Impermeant in salt form	Argon ion (488)	Acetoxymonoester group renders Fluo-4 cell permeant
Mag-Fura	Mg^{2+}	Impermeant in salt form	N_2 (337)	See note for Fluo-4
Mag-Fluo-4	Mg^{2+}	Impermeant in salt form	Argon Ion (488)	See note for Fluo-4
SBFI	Na^+	Impermeant in salt form	N_2 (337), Nd:YAG (335)	See note for Fluo-4 regarding permeability. Na^+ and K^+ are cross-reactive
PBFI	K^+	Impermeant in salt form	N_2 (337), Nd:YAG (335)	See note for Fluo-4 regarding permeability. Na^+ and K^+ are cross-reactive

[a] Some ion probes show cross-reactivity with other ions. The target ion listed typically has the highest.

Table 9.5 Standards for flow cytometry, microscopy, and cell separations

Application	Standard	Examples
Intensity (Bright vs. Dim)	Beads stained as bright, medium, or dim for several fluorophores	Right Reference Standard (BL)
Intensity (Relative Scale)	Beads stained in linear or log-scale increments of each other (e.g., 100%, 50%, 10%, 1%, 0.1%, etc.)	LinearFlow beads (IV), Rainbow Linear Calibration Particles (SP)
Intensity (MESF)	Beads calibrated with known MESF (as compared to solutions of same MESF)	Quantum MESF (BL)
Bead Size	Several beads of monodisperse size in each set	Flow Cytometry Size Calibration Kit (IV), Particle Size Standard Kit (SP), Size Calibration Standards Kit (BL)
Alignment (Flow Cytometer)	Broad-spectrum beads to align multiple lasers and several detectors	Full Spectrum (BL), Rainbow Fluorescent Particles (SP), Fluorescent Alignment Particles (SP), Alignflow (IV)
Compensation	Beads stained with different fluorophores (at similar levels to stained cells) for easy compensation. May be spectrally matched beads or antibody-binding type.	Compbead (BD), SPHERO COMPtrol (SP), SPHERO Easycomp (SP), Simply Cellular Compensation Standard (BL)
Focus (Microscopy)	Bead containing an inner core dye and an outer layer of differing dye. Used in confocal systems	FocalCheck (IV)
Absolute Counting	Beads calibrated in units of beads ml^{-1}. Can be used to calibrate flow cytometer, or added to samples to get absolute counts without hemacytometer	Absolute Count Standard (BL), AccuCount Particles (SP), CountBright Absolute Counting Beads (IV)

(continued)

Table 9.5 (*Continued*)

Application	Standard	Examples
Antibody-Binding Capacity (Flow Cytometry and Cell Separations)	Bead set with increasing numbers of secondary antibody on surface	Quantum Simply Cellular (BL)
Biotinylated Beads (Microscopy, Flow Cytometry, Cell Separation)	Biotin-coated beads to test conjugation to avidin-type probes or avidin-coated surfaces. Available with dyed beads	Biotin-Labeled Microspheres (IV), Proactive Biotin Beads (BL)
Streptavidin Beads (Microscopy, Flow Cytometry, Cell Separation)	Streptavidin beads to test conjugation to biotin-type probes or binding to biotinylated surfaces. Available with dyed beads	Proactive Streptavidin Beads (SP), Streptavidin-Labeled Microspheres (IV).

BL = Bangs Laboratories; BD = Becton-Dickinson Biosciences; IV = Invitrogen/Molecular Probes; SP = Spherotech. Note: Examples are not exhaustive; other vendors with similar products will yield acceptable results.

Table 9.6 Amine-reactive and thiol-reactive labels

Functional group on protein	Reactive group on dye	Structure[a]	
Primary Amines ($-NH_2$)	Isothiocyanate	$RN=C=S$	Primary amines located on lysine residues and N-terminus
	Succinimidyl Ester		
	Sulfonyl Chloride		
Thiol ($-SH$)	Maleimide		pH controlled to selectively label thiols instead of amines

[a] R = Fluorescent molecule.

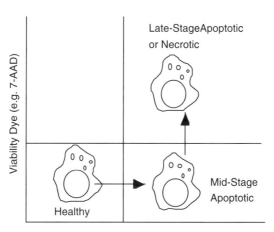

Annexin-V Fluorescence

Figure 9.4 Flow cytometer plot showing regions of apoptosis. Healthy cells will not stain brightly for fluorophore-conjugated Annexin-V. Dead cells (and late-stage apoptotic) will stain for both Annexin-V and the viability dye (e.g., 7-aminoactinomycin D, 7-AAD). Mid-stage apoptotic cells will stain positively for Annexin-V, but have intact membranes and therefore will be negative for the viability dye. Note: If calcein-AM is used as a dye, then healthy and mid-stage apoptotic cells be brightly stained and dead and late-stage apoptotic cells will be dimly stained

staining should be done in buffer containing sufficient Ca^{2+}. Many vendors supply Annexin staining buffer, which is typically 0.14 M NaCl, 2.5 mM $CaCl_2$ in pH 7.4 HEPES buffer. The relatively high calcium concentration has been reported to be toxic to some cell lines [7]. If this possible toxicity is a concern, the Ca^{2+} concentration can be titrated. When performing apoptosis assays, it is helpful to have both a positive and a negative control. The positive control will be a sample incubated with an agent known to induce apoptosis. Staurosporine and camptothecin are two agents that work well for this type of positive control. Both agents induce apoptosis via the caspase-9 pathway. Certain antibodies for the Fas receptor (CD95, which induces apoptosis via the caspase-8 pathway) can be used as well. The negative control can be a sample without the induction agent, or it can be an induced sample incubated with a pan-caspase inhibitor to inhibit apoptosis.

Reagents needed: Annexin-binding buffer, Annexin-V conjugated to a fluorophore, a DNA viability dye such as 7-AAD or propidium iodide, an apoptosis inhibitor, such as the pan-caspase inhibitorz-DEVD-FMK (benzyloxycarbonyl-valine-alanine-aspartic acid-fluoromethyl ketone).

Note: Fixed cells cannot be analyzed by this method.

Preparing the Positive (Apoptotic) and Negative Control

1. Prepare cells as a suspension (see Chapter 3) at a volume of 10^5-10^6 cells ml. Cells can be kept in culture medium during the induction period.
2. Incubate cells with either buffer containing 1 µM staurosporine or $0.5-1.0 \, \mu g \, ml^{-1}$ of an anti-CD95 (anti-Fas) antibody that causes apoptosis. For the negative (not induced) control, see Step 3.

    Note: Not all anti-CD95 antibodies induce apoptosis. Check with manufacturer. The functional anti-CD95 antibody clone EOS9.1 from eBioscience works well with a variety of cell lines.

    Note: Staurosporine should be handled with care to avoid accidental exposure.
3. For the negative control, suspend cells as in step 1 and incubate in medium until the analysis (step 5).

Preparing an Inhibited Control

4. Suspend a third cell sample in culture medium and add the induction agent (as in step 2). Immediately add 20 µM of the z-VAD-FMK inhibitor and incubate cells in medium until analysis (Step 5).

Analysis by Flow Cytometry or Fluorescence Microscopy

5. Before analysis (and after the required induction time), centrifuge cells and resuspend in Annexin-binding buffer at a concentration of 10^5–10^6 cells ml^{-1} (this concentration can be adjusted as needed).

6. Add one test of fluorophore conjugated Annexin-V and 0.01 mg ml^{-1} (final concentration) of 7-AAD (or similar). Incubate for 5–20 minutes.

7. For microscopy, place 20–30 µl drop of cell suspension on a cover slip and allow the cells to settle to surface. Acquire white-light and fluorescence images (for both probes) for each field of view.

8. For flow cytometry, apoptotic cells may appear as a different population from healthy cells on a dot plot for forward and side scatter. Gate both populations, if observed. Fluorescence for healthy and apoptotic cells can be gated as shown in Figure 9.4.

   Note: Uninduced control samples will have a small percentage (<5–10%) of dead and apoptotic cells, respectively. Inhibited controls should have ≪5% of apoptotic cells. Annexin-V staining will increase over several hours.

PROTOCOL 9.5: APOPTOSIS DETECTION USING FLUOROGENIC CASPASE PROBES

Annexin-V binding to phosphatidyl serine is a marker of mid- to late-stage apoptosis. It is desirable in many cases to detect apoptosis at an earlier stage, or to detect the caspase family of proteins that are responsible for apoptosis. The main group of caspase probes takes advantage of the so-called specificity of caspases for certain amino-acid sequences (Table 9.3). All caspase proteins cleave at aspartic acid residues, and are widely assumed to have a high affinity for only a certain sequences. However, it has been shown [8] that caspases cleave other sequences with reasonable affinity. For example, the caspase 3 "specific" tetrapeptide sequence DEVD (aspartic acid, glutamic acid, valine, aspartic acid) is also cleaved by caspase 7. Care must therefore be taken when assigning fluorescence signals to the activity of a particular caspase.

One must also consider which caspase to "target," even though more than one caspase will produce a percentage of the total signal. For example, receptor-mediated apoptosis will activate the caspase 8 initiator, while cytotoxic damage will activate the caspase 9 initiator. Regardless of the mechanism of induction, caspases 3, 6, and 7 will all be activated as effector or executioner caspases. A caspase 3 probe will produce a signal slightly later than a caspase 8 or 9 probe, since caspase 3 is activated after the initiator caspases. One class of caspase probes, fluorescent inhibitors of caspases (FLICAs), can be incubated with apoptotic cells. These probes bind irreversibly and produce a signal related to the total number of active target caspases. FLICAs require a washing step. They are available for several different tetrapeptide sequences (Table 9.3). An alternate approach is to use a fluorogenic reagent. A rhodamine 110 dye [9] modified with two amino-acid groups or sequences is cell permeant, but nonfluorescent in its intact state. Once cells become caspase active, the amino-acid groups or sequences are cleaved in a two-step process to produce the free rhodamine 110 dye, which is retained in the cell. The benefit of this approach over FLICAs is that washing is not necessary. Unlike FLICA assays, the fluorescence intensity is not directly proportional to the total number of activated caspases in the cell, as the fluorescence increases both with the number of caspases and the incubation time. Fluorogenic probes containing only aspartic acid (as a pan-caspase probe), DEVD (targeting caspase 3), IETD (caspase 8), and LEHD (caspase 9) are commercially available. The following protocol is for apoptosis measurement using the pan-caspase rhodamine probe bis(L-aspartic acid) rhodamine 110.

Reagents needed: Induction and inhibition agents (see Protocol 9.4), phenol-red-free medium (if continuous observation is desired), caspase probe (see Table 9.7 for examples of other fluorogenic probes).

1. See Protocol 9.4 for preparing apoptotic control, noninduced control, and inhibited control.
2. For all samples and controls, incubate cells with 10 µM bis(L-aspartic acid) rhodamine 110 as soon as samples are induced.
3. For end-point measurements, incubated cells until the desired time (e.g., 3 hours), wash, and resuspend in phosphate-buffered saline (pH = 7.4). Then mount on a microscope slide or prepare for flow cytometry.
4. For continuous measurements (e.g., intensity vs. time of incubation), cells can be placed in a flow cell (Chapter 7) and maintained in phenol-red-free medium and observed by microscopy. Alternatively, they can be placed in a microscope chamber (Chapter 4) in phenol-red-free medium and observed by microscopy. While light and fluorescence images should be obtained, only one field of view

Table 9.7   Examples of fluorogenic probes

Probe type	Application	Target	Notes
Rhodamine-110 (R110) Fluorogenic Protease Probes[a]	Apoptosis, other protease activation	Caspases, other proteases	R110 amine groups modified with amino acids or peptides recognized by target proteins. Cleavage of both groups produces fluorophore
Nile-Blue Analog Protease Probes [19]	Protease activity	Specific protease of interest	Red-excited, red-fluorescent probe
NucView Protease Probes	Apoptosis	Caspases	Caspases cleave specific peptide sequence, releasing a DNA-binding dye that enters the nucleus and becomes fluorescent
Calcein-AM and Other AM-Modified Dyes.	Viability, loading of various impermeant dyes into cells	Nonspecific esterases in the cell	AM group provides permeability as well as disrupting fluorescence. Useful mechanism for loading dyes into living cells
FRET-Based Protease Probes	Protease activity in cells	Specific proteases of interest	Cells must be transfected to produce FRET probe. Fluorescent proteins coupled by peptide sequence recognized by protease
Fluorescein Diphosphate (FDP)		Alkaline phosphatase	Produces fluorescein upon cleavage
Propidium Iodide (PI)	Viability, DNA labeling	DNA	While not technically fluorogenic, the fluorescence of unbound PI is much lower than when bound to DNA

[a] Other fluorophores are available for this type of caspase probe.

can be observed if the same set of cells is to be observed over time (Figure 9.5).

5. For continuous measurements by flow cytometry, aspirate suspended cells from the sample chamber (Step 4), wash and resuspend,

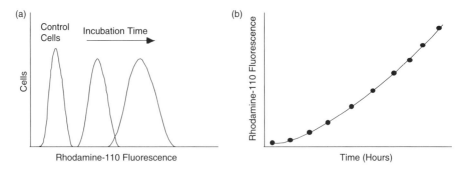

Figure 9.5   (a) Histogram of rhodamine 110 fluorescence produced from fluorogenic caspase probes. The median fluorescence of apoptotic cells will increase with time. (b) The fluorescence intensity of the apoptotic population increases over time, but will ultimately decrease as cells dye and caspase activity lessens at the end stages of apoptosis

and measure the cell sample. A new aliquot of cells must be taken for each time point. For measurement of the same cells over time, see step 4.

9.5   SURFACE MODIFICATIONS FOR CELL ANALYSIS (CHAPTERS 5 AND 7)

PROTOCOL 9.6: COVALENT LINKAGE OF PROTEINS (NONANTIBODY) TO GLASS BY MICROCONTACT IMPRINTING

The permanent, covalent attachment of proteins to glass requires modification of the terminal SiOH groups found on the glass surface (Figure 9.6). Glass by itself is not highly reactive, especially when one considers that the main methods to conjugate antibodies and other proteins to surfaces involve thiol/sulfhydryl (SH), carboxylic acid (COOH), and amine (NH_2) functional groups. Protocol 9.7 deals with the conjugation chemistry required to attach antibody proteins to glass covalently. When conjugating proteins to glass, there are several strategies that take advantage of residual carboxylic acid or amine groups. Preparing a glass surface with a reagent such as 3-amino-propyltriethoxysilane allows for the protein to be coupled directly to the surface with minimal preparation. As with any conjugation approach, it is best to check the binding or activity of the immobilized

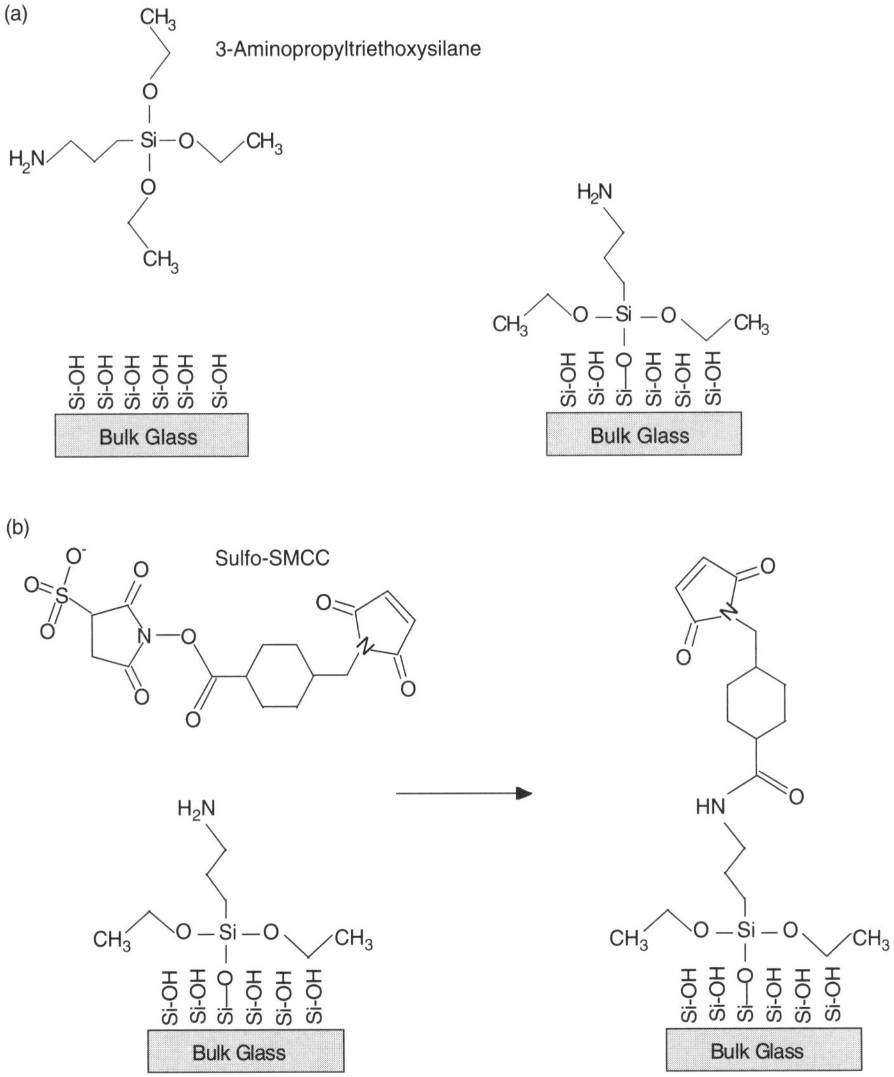

Figure 9.6 (a) Modification of glass to produce an amine-covered surface; (b) preparation of a maleimide surface

protein before use in an analytical system to ensure the protein was not rendered inactive.

When conjugating glass surfaces, the cleanliness of the starting materials will dictate the success of the experiment. While cleaning glass with piranha solution is commonly used, the violent nature of this cleaning mixture poses a serious safety risk. The cleaning method outlined in this protocol has been used for a variety of glass-surface applications. Alternatively, a plasma cleaner may be used to prepare the surfaces for conjugation.

Reagents/materials needed: *Clean* glass surfaces, 3-aminopropyl-triethoxysilane, pure acetone for 3-aminopropyltriethoxysilane. >95% pure heptane, isopropanol, and methanol (for cleaning slides). Buffer (phosphate-buffered saline, PBS, containing EDTA: 50 mM phosphate buffer containing 10 mM EDTA (150 mM NaCl, pH = 7.2)), (Bis-[sulfosuccinimidyl] suberate), Tris, poly(dimethylsiloxane), either patterned or flat PDMS slab.

Preparation of the Glass Surface Using an Amine Functional Group

1. To clean glass surfaces, immerse slides in heptane for 30 seconds. Rinse glass with isopropanol three times followed by three washes with high-grade methanol or acetone.
   Note: Other glass cleaning methods can be used.
2. Dilute 3-aminopropyltriethoxysilane to 2% (v/v) in high-grade, anhydrous acetone.
3. Either cover the glass surfaces with the dilute 3-aminopropyl-triethoxysilane or immerse them. In both cases, react for 30–60 seconds at room temperature.
4. Rinse three times with anhydrous acetone, air dry in clean container.
5. Modified glass surfaces can be stored until needed in a clean container.

Conjugating Proteins to Amine-Modified Glass via Lysine Residues (Method 1)

In this approach, a homobifunctional cross-linker, (bis[sulfosuccinimidyl] suberate), is used to conjugate the amine on the glass surface with amines from lysine residues on the protein [10]. Since this cross-linker contains two N-hydroxysuccinimide ester (NHS) groups, there is a possibility that the proteins may, essentially, polymerize in solution, as the bis[sulfosuc-cinimidyl] suberate may cross-link two proteins. As a measure to prevent this, the reaction is carried out on a poly(dimethylsiloxane) (PDMS) surface [10]. The PDMS surface can be flat or patterned (in order to deliver the protein in a pattern on the glass. The PDMS surface is then brought into contact to the amine-modified glass to complete the reaction.

Spatial control of the protein deposition is possible if the PDMS surface is patterned.

6. The protein concentration will vary by application, but a range of 8–10 mg ml^{-1} will work in most cases and can be used as a starting point for optimization. Prepare in PBS buffer with EDTA.
7. Coat the protein solution on the surface of a slab or patterned piece of PDMS. Remove excess liquid by blotting or drying. Add the bis-[sulfosuccinimidyl] suberate cross-linker and react on the PDMS surface for 30 minutes at room temperature.
8. Add 1 M Tris to quench for 15 minutes.
9. Remove excess liquid from the PDMS surface by blowing the surface with dry nitrogen rapidly.
10. Immediately bring the PDMS surface (with reacted protein) into contact with the amine-modified glass for 5 minutes.
11. Remove PDMS gently. Protein-modified glass should be used as soon as possible.

PROTOCOL 9.7: COVALENT LINKAGE OF ANTIBODIES TO GLASS

Protocol 9.6 dealt with the attachment of proteins to modified glass. These methods can be used, to some degree of success, with antibodies as well. However, the danger associated with glass conjugation for any protein is that the binding site or active site may be sterically hindered or rendered inactive. In antibody conjugation, one can take advantage of the disulfide bridges already present. The disulfide bridges that join the two heavy chains are cleaved (Figure 9.6), leaving two monovalent antibody halves that can each bind one analyte. Alternately, the antibody can be modified with additional sulfhydryl groups. In this method, the antibody is left intact and remains divalent. Both methods must be carefully controlled to minimize loss of antibody functionality. Once the antibody is prepared with thiol groups, it can readily react with maleimide groups attached to glass surfaces. This protocol is modified from protocols from Pierce Biotechnology [11]. Greg Hermanson's book, *Bioconjugate Techniques* [12], is also a recommended reference when performing conjugation reactions.

Reagents/materials needed: *Clean* glass surfaces, 3-aminopropyltriethoxysilane, pure acetone for 3-aminopropyltriethoxysilane, sulfosuccinimidyl-4-(N-maleimidomethyl)cyclohexane-1-carboxylate (Sulfo-SMCC), Traut's Reagent (2-Iminothiolane-HCl), 2-Mercaptoethylamine (2-MEA), size-exclusion columns or dialysis cartridges, buffer (phosphate-buffered saline, PBS, containing EDTA: 50 mM phosphate buffer containing 10 mM EDTA (150 mM NaCl, pH = 7.2)).

1. See Protocol 9.6, steps 1–5 for the preparation of glass surfaces with amine groups via 3-aminopropyltriethoxysilane. These glass surfaces may be prepared and stored.
2. To produce maleimide surfaces [11], mix a 2 mg ml^{-1} solution of sulfo-SMCC in PBS buffer containing EDTA. This reagent contains an amine-reactive group (N-hydroxysuccinimide, or NHS) and the thiol-reactive maleimide group. Use immediately.
3. Cover amine-modified glass surface with sulfo-SMCC solution and incubate for 1 hour at 20 °C. Rinse surfaces with the same phosphate–EDTA buffer. Maleimide surfaces may be stored at 4 °C in a desiccator (the surfaces must remain dry until needed).

Preparation of Intact Antibodies (skip to step 9 if antibodies will be reduced instead)

4. Prepare an 8–9 mg ml^{-1} solution of antibody in approximately 0.5 ml of PBS buffer containing EDTA (pH = 8.0).
5. Add Traut's reagent to a final concentration of 0.7–0.8 mM to the antibody solution (the added volume should be small, 25–50 μl).
6. Incubate 30–60 minutes at room temperature.
7. Separate the antibody from the remaining reagent using either a size-exclusion column or microdialysis cartridge. If size exclusion is used, collect fractions and save fractions with absorbances at 280 nm.
8. Immediately couple to glass surface (see Step 13).

Preparation of Reduced Antibodies

9. Prepare an 8–9 mg ml^{-1} solution of antibody in approximately 0.5 ml of PBS buffer containing EDTA (pH = 7.2).
10. Add 2-MEA to a final concentration of 0.05 M to the antibody solution (the added volume should be small, 25–50 μl).
11. Incubate at 37 °C for 90 minutes.
12. Separate the antibody as in step 7, then proceed to step 13 immediately.

Conjugation of antibody to Maleimide Surface

13. Dilute the antibodies produced in steps 4–8 or 9–12 in PBS buffer containing EDTA (pH = 7.2) so that the volume is sufficient to cover the amine-modified glass surface as desired.
14. Incubate at room temperature for 2 hours. Longer incubation times may be needed in some cases.
15. Rinse with PBS buffer containing EDTA.

While surfaces may be stored desiccated or in buffer containing dilute (0.02–0.1% sodium azide) at 4 °C, it is recommended that antibody-covered surfaces be used as soon as is practical.

PROTOCOL 9.8: NONCOVALENT ATTACHMENT OF ANTIBODIES TO GLASS #1

This method uses a sandwich approach to producing a coating of neutravidin, streptavidin, or avidin that can then be used with a biotinylated capture molecule (Figure 9.7). Using biotin–avidin chemistry, surfaces can be produced and stored that can accept a variety of biotinylated capture molecules. This protocol is derived from the work from Flavell's laboratory [13] and has been modified for use in cell capture [14]. The method is rapid, uses minimal reagents, and is performed at room temperature in aqueous solution.

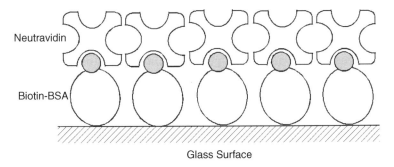

Glass Surface

Figure 9.7 Noncovalent attachment of biotinylated capture molecules for cell separation. A layer of biotinylated bovine serum albumin is deposited onto cleaned glass. Neutravidin, streptavidin, or avidin is then added, forming a layer that can accept any biotin-conjugated capture molecule, such as a biotinylated antibody

Reagents/Buffers:
T50 Buffer: 10 mM Tris-HCl, pH 8.0, 50 mM NaCl; neutravidin (0.2 mg ml^{-1} in T50 buffer); biotinylated bovine serum albumin (Biotin-BSA, 1 mg ml^{-1} in T50 buffer) (avidin or streptavidin may be substituted for neutravidin; however, neutravidin has low nonspecific binding characteristics and works well in cell separations); glass slides, capillaries, or microfluidic chips for noncovalent attachment. For glass slides, electrostatically coated slides work well. Glass capillaries have been used without modification.

1. Wash coverage surfaces with T50 buffer (5–10 minutes). For capillary or microfluidic chip coating, use a syringe and aspirate all fluids into the channel, rather than forcing fluid through (doing so will reduce the formation of bubbles).
2. Cover surface with 1 mg ml^{-1} biotin-BSA solution for 10–45 minutes (smaller channels may require the longer incubation times).
3. Wash with T50 buffer, then cover surface with 0.2 mg ml^{-1} neutravidin solution.

Cover surface with 0.2 mg ml^{-1} neutravidin solution and incubate for 10–45 minutes).

4. Flush twice with T50 buffer, then once with deionized water.
5. Gently flow clean air or nitrogen gas through the channel or over the surface to dry.
6. Modified surfaces can be stored in a 4 °C refrigerator until needed.
7. Before use, a biotinylated capture molecule (antibody, aptamer, etc.) should be added to the surface/channel and incubated for ten minutes before rinsing with phosphate buffered saline (pH = 7.4) containing 3% bovine serum albumin (not biotinylated).

   This coating method can be used to produce affinity surfaces for cell capture that can withstand high shear flow and exhibit low nonspecific binding.

PROTOCOL 9.9: NONCOVALENT ATTACHMENT OF ANTIBODIES TO GLASS OR PDMS #2

To coat PDMS surfaces, or glass surfaces when *low shear flow* will be used, Protocol 9.8 can be modified to omit step 2 (biotin-BSA). Neutravidin is deposited onto glass or PDMS, then stored until needed [15]. The biotinylated capture molecule is then added before use.

PROTOCOL 9.10: BLOCKING ENDOGENOUS BIOTIN

Biotin is present at low levels in many cells. Given that biotin–avidin chemistry is one of the cornerstones of cell labeling, endogenous biotin may impose a systematic error when staining, particularly with sensitive techniques such as single-molecule fluorescence microscopy. To block endogenous biotin, avidin or a similar protein is added to the sample to bind to the biotin present in the cells. To block the free binding sites of avidin (four binding sites total per protein), excess, free biotin is then added, followed by a final wash. The end result is that all endogenous biotin is blocked by avidin, which is then rendered incapable of binding to any additional biotin. Subsequent staining with biotinylated and streptavidin-conjugated probes can then continue. This process may be performed using reconstituted powdered skim milk as the biotin source and egg white as the avidin source. However, these natural materials may add more impurities and artifacts than relatively pure solutions of biotin and avidin.

Materials needed: T50 Buffer (10 mM Tris-HCl, pH 8.0, 50 mM NaCl), biotin solution (0.4 mg ml^{-1}) in T50 Buffer, neutravidin (avidin and streptavidin may be used) at a concentration of 0.1 mg ml^{-1} in T50 buffer, 3% solution of bovine serum albumin in T50 buffer (w/v).

1. Wash cells with 3% bovine serum albumin solution, followed by three washes with T50 buffer (suspended cells should be centrifuged and resuspended three times).
2. Incubate cells at room temperature for 10–15 minutes with the neutravidin solution.
3. Wash three times with T50 buffer.
4. Incubate cells with biotin solution for 30–40 minutes (longer times may be needed in some cases).
5. Wash three times with T50 buffer.
6. Proceed with subsequent biotinylated and neutravidin-linked probe staining.

Note: Blocking protein adsorption on glass, PDMS, or other surfaces is essential for many cell analyses. In particular, sensitive measurements of cells and their contents require suppression of surface adhesion. In cell separations, blocking protein adhesion will reduce nonspecific binding of background cells to the affinity surface. There are several blocking agents available commercially; many of them contain proprietary mixtures. Effective blocking in many cases can be performed using a 3–5% solution of bovine serum albumin in either phosphate-buffered saline or T50 buffer.

9.6 FLOW CYTOMETRY AND CELL SEPARATIONS (CHAPTERS 5 AND 6)

PROTOCOL 9.11: CELL CYCLE MEASUREMENTS BY FLOW CYTOMETRY

Cell-cycle measurements are important in cell proliferation, cancer, and apoptosis. One of the simplest methods of cell-cycle measurement is to use a fluorescence DNA stain and measure cell fluorescence by flow cytometry. While cell-cycle measurements by fluorescence microscopy are possible, flow cytometry will yield better statistical data, especially when looking for deviations from the normal cell cycle.

When conducting cell-cycle analyses, the choice of DNA stain and standards can affect the outcome of the experiment considerably. The ideal DNA stain should be cell permeant and selective only for DNA (and not bind to RNA). Unfortunately, dyes that come close to meeting these requirements, such as Hoechst 33342 – are largely excited by UV light and not suitable for many of the smaller, bench-top flow cytometers. Dyes excited at the ubiquitous 488 nm line and the increasingly popular 532 nm line include propidium iodide, acridine orange, and 7-AAD (see Table 9.2). These dyes are not DNA selective, and many are impermeant. To overcome these two obstacles, cells are often fixed and then treated with RNAse to remove endogenous RNA.

Since many immortalized cell lines are aneuploid, a stable reference material is required for some cell-cycle measurements. Chicken and trout erythrocytes, which are both nucleated and senescent, serve as suitable standards for cell-cycle measurements, as well as any stable, diploid cell line. Fluorescence is acquired in linear mode on the flow cytometer, since the DNA difference in most cases is a factor of two to four (see Figure 9.8). The following protocol makes use of propidium iodide, one of the most common DNA stains, although other dyes can be substituted with minimal modification. The cells are first fixed and permeabilized using ethanol, followed by removal of RNA and subsequent staining and analysis.

Materials needed: 90% ethanol, RNase (DNase-free), propidium iodide (1–5 μg ml^{-1} in pH = 7.4 phosphate buffered saline containing 0.1% Triton-X).

1. Place cells in suspension at a concentration of 10^5–10^6 cells ml^{-1} in phosphate-buffered saline (note: wash cells at least once to remove residual protein from the culture medium).

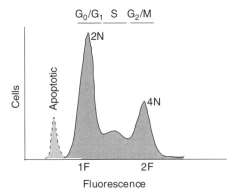

Figure 9.8 Conceptual depiction of the linear fluorescence histogram of DNA stain fluorescence and cell number. Diploid (2N) cells in the G_0/G_1 phase will have a base fluorescence (1F) that is half the fluorescence intensity of tetraploid (4N) cells in the G_2/M phases of the cell cycle. Cells in the synthesis (S) phase of DNA replication will have a peak fluorescence between the 1F and 2F intensities. Apoptotic cells undergoing chromatin fragmentation will show a peak with intensity less than 1F

2. Centrifuge cells for 4–5 minutes to create a pellet. Discard supernatant.
3. Add cold 90% ethanol dropwise while stirring until the sample is resuspended in 0.5–1 ml of 90% ethanol. The cell sample can be placed in a $-20\,°C$ freezer overnight. A fixation time of at least 2 hours is needed before analysis; samples can be stored for one year.
4. Centrifuge samples and resuspend in phosphate buffered saline three times. It is critical that as much of the ethanol is removed as possible.
5. Add RNase to a final concentration of $0.2\,mg\,ml^{-1}$ in the propidium iodide solution. Incubate for 20 minutes at $37\,°C$ before analyzing by flow cytometry.

   Note: After adding ethanol cells may clump. Pipette up and down in sample tube to disrupt clumps before staining or analysis.

PROTOCOL 9.12: ANTIGEN DENSITY MEASUREMENTS IN FLOW CYTOMETRY

The measurement of antigen density is critical for a variety of analyses, especially when modeling or predicting the outcome of a cell-affinity experiment. The typical method of determining antigen density (i.e., antigen molecules per square micrometer of the cell surface) requires two measurements. First, cell surface area must be determined by microscopy or flow cytometry. Second, the total number of antigens on the cell must be determined.

The main drawback, as pointed out in *Practical Flow Cytometry* [16], is that measurement of mean cell size is prone to systematic error. Flow cytometry measurements based on light scatter require calibration, and measurement by microscopy is typically performed by assuming the cell is a sphere. The diameter of a larger number of cells is measured by microscopy and the surface area calculated. Given the natural variation in antigen density and cell size, this last assumption may be a negligible error, but this is not guaranteed. Differences in cell shape and morphology will more often than not deviate from the spherical model. However, there are few alternatives to accurate cell surface-area measurement; therefore, antigen density values obtained by measuring cell size should be considered to be an estimated value. The benefit of measuring cell size by microscopy is that additional reagents are not needed.

The cell size can be estimated by measuring cells by flow cytometry, where the count number is significantly higher. However, for accurate measurements, the instrument should be calibrated. Calibrating a flow cytometer using beads of known size is one method that is commonly used. The drawback of using this approach is that the bead refractive index (which affects forward-scatter intensity) will be different from the cell refractive index. Again, a systematic error is incurred, affecting the accuracy of the measurement. In most cases, however, the variation in the naturally occurring antigen density and cell surface area will overshadow the systematic errors discussed.

Equipment needed: Flow cytometer with detection optics for the fluorophore-conjugated antibodies used, light microscope with CCD camera (for cell size measurements).

Reagents needed: Fluorophore-conjugated antibody (or primary antibody and fluorophore-conjugated secondary antibody), calibration beads.

Note: Two different calibration strategies can be used. In the first strategy, a set of beads determines fluorescence intensity as a function of the number of fluorophores. The second bead set determines the number of fluorophores per antibody. This approach allows for greater flexibility in the lab at increased cost and time of analysis.

The second strategy uses a set of beads to calibrate the fluorescence intensity as a function of the number of antibodies bound to a cell (effectively combining both steps of the first method into one). The approach is less flexible (i.e., the beads are purchased to match the primary antibody isotype, but is less expensive and requires less analysis time. The second approach is presented here.

The calibration beads used in this analysis, Quantum Simply Cellular Antibody Binding Capacity beads, are available from Bangs Laboratories, although other, suitable products can be used as well. This bead set contains a series of beads, each coated with a known number of secondary antibodies (in this example anti-mouse IgG). The benefit of using these beads for measuring antibody-binding capacity is that the fluorophore-conjugated primary antibody may be used directly. Several beads, including a blank set without secondary antibodies, are used to generate a calibration curve of antibody-binding capacity vs. fluorescence. To obtain antigen density, microscope images of the cells under investigation can be made, and the mean volume of the cells estimated by measuring the cell diameter. The antibodies coated on the beads must bind the epitope of the primary antibodies under investigation.

1. While many protocols recommend labeling all the beads simultaneously, in some cases it may be better to stain each bead type individually to ensure that two beads that appear to have similar intensity on a particular instrument will not overlap.
2. Use one drop of each bead type for $100\,\mu$l of buffer (3% w/v bovine serum albumin in phosphate-buffered saline, pH $= 7.4$ works well).

   Add fluorophore-conjugated primary antibody to saturate each bead sample; typically add enough antibody to stain 1×10^6 to 5×10^6 cells.

   Note: under-saturating the beads will produce a nonlinearity with the higher antibody-binding-capacity beads.
3. Incubate at room temperature for 20–30 minutes (longer incubation times may be needed for some antibodies). Protect samples from light.
4. Centrifuge twice for 4 minutes at $1100 \times g$, each time discarding the supernatant and replacing with an equivalent volume of buffer.
5. Resuspend in $500\,\mu$l of buffer and analyze on the flow cytometer.
6. The flow cytometer settings should be adjusted using the blank bead set. The fluorescence measurements used for this calibration may differ from what is normally used when measuring the cells of interest. The detector voltage or gain settings should be set so that

the blank beads fall within the first decade of signal intensity on a log scale (see Figure 6.7 in Chapter 6).

7. When analyzing data, make sure to gate only the single-bead events, and exclude double occupancies (Figure 6.5 in Chapter 6).

8. After acquiring data for the beads, each cell line tested with the primary antibody should then be measured using the *same settings as the bead analysis*, not the settings normally used for cell counting/identification.

9. Five bead populations of increasing intensity should be detected. Each population (see Figure 9.9) should be gated and the median fluorescence intensity recorded. A plot of fluorescence intensity vs. antibody binding capacity can then be used to determine the antibody-binding capacity of the cells.

10. The antibody-binding capacity of the blank beads should be determined, and subtracted from all cell samples.

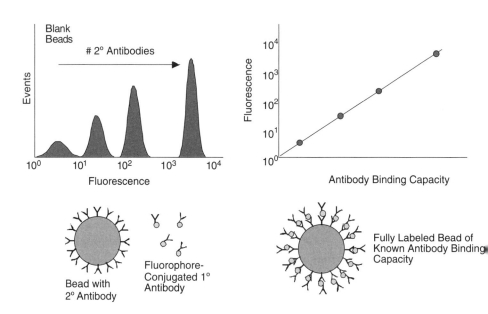

Figure 9.9 Conceptual diagram of the fluorescence intensity histograms (upper left) of Quantum Simply Cellular Beads containing increasing numbers of secondary antibodies stained with the primary antibody of interest. The blank (unstained) bead population should fall in the first decade of the log scale of fluorescence. A calibration curve of bead antibody-binding capacity and fluorescence is used to determine the antibody-binding capacity of subsequent cell samples (top right). Each bead contains a known number of secondary antibodies (bottom left and right) that are used to calibrate the number of antibodies stained on a cell

PROTOCOL 9.13: ANTIGEN DENSITY MEASUREMENTS USING FLUORESCENCE CORRELATION SPECTROSCOPY

While flow cytometry is a routinely used method to produce antigen density measurements, there are some limitations to the technique. Antigen density measurements in flow cytometry require the use of calibrated antibody-binding capacity beads, followed by a measurement of cell size. In some cases, for example when new molecular probes are developed, a bead set with secondary antibodies is not available. The size measurement is, as noted in Protocol 9.12, also prone to systematic error. Antigen density can also be measured using fluorescence correlation spectroscopy (FCS) [17] to probe a membrane surface. In this case, the FCS signal is used to determine antigen density directly (Figure 9.10). FCS has been used for a variety of cell-transport studies, as well as sensitive detection of intracellular molecules. By focusing a confocal microscope laser onto the membrane surface, the number of fluorescent probes can be determined by FCS. If the spot size is accurately known (and can be calibrated easily), then the number of antigens per μm^2 is directly measured [18].

Equipment needed: Microscopy capable of confocal detection (scanning is not required) and fluorescence correlation spectroscopy detection.

Reagents needed: Antibody or other surface marker label. This primary antibody or marker label may be conjugated to a fluorophore. An alternative labeling strategy is to use a biotinylated antibody and a neutravidin-conjugated fluorophore. In this approach, endogenous biotin present in the sample must be blocked with avidin (Protocol 9.10). A third

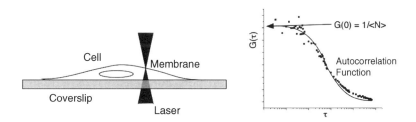

Figure 9.10 Fluorescence Correlation Spectroscopy (FCS) measurements of antigen density. A laser is focused onto the membrane of a stained cell and autocorrelation measurements initiated. The autocorrelation function, $G(\tau)$, will be generated for fluorescent antibodies on the membrane surface. The amplitude of this function, $G(0)$, is inversely proportional to the number of fluorophores measured. If the beam area at the focal point is measured, antigen density is determined

labeling strategy is to use an unlabeled primary antibody and a labeled secondary antibody.

Note: Adherent cells should be grown on glass cover slips or surface-modified dishes (Chapter 4).

1. Stain cells using the primary antibody in the presence of 3–5% bovine serum albumin (%w/v) for the appropriate time (10–30 minutes typical).
2. Wash cells (if adherent, gently wash with flowing buffer; if suspended, centrifuge and resuspend in buffer or phenol red free medium).
3. If primary antibody or surface marker label is conjugated to a fluorophore, proceed to step 5.
4. Add secondary detection reagent, incubate, and wash as in steps 1–2.
5. Place cell sample on microscope. Position a cell in the laser-beam path with the laser shuttered. Laser power should be set to minimum value where fluorescence can be detected to avoid photobleaching (5–200 μW typical, the value should be optimized depending on wavelength, dye, cell, etc.).
6. Open laser shutter and focus the beam on the cell membrane. Acquire fluorescence/FCS data.

   Note: focusing above or below the cell membrane will produce reduced autocorrelation and signal.
7. Repeat process on second position of cell, then repeat steps with a new cell. Measure at least 50 cells to obtain representative data.
8. Data can be fit using the following function for membrane-bound probes:

$$G(\tau) = \frac{1}{\langle N \rangle} \left(1 + \Gamma \frac{\tau^{\alpha}}{w^2} \right)^{-1}$$

where $\langle N \rangle$ is the average number of fluorescent molecules in the probe volume, Γ is the transport coefficient, α is a nonlinear parameter ($0 < \alpha < 1$) and w is the $1/e$ radius of the laser-beam focus. The laser spot radius can be determined by measuring FCS curves of freely diffusing dyes of known diffusion coefficients and fitting them to a 2- or 3D diffusion model.

Note: The ratio of fluorophores to antibodies/probes must be known and used to correct for the number of antibodies measured (not the number of fluorophores).

9. The antigen density can be calculated by this method as the number of molecules (derived from $\langle N \rangle$) detected over the spot size with area $A = \pi w^2$, where, again, w is the laser spot/confocal detection volume radius.

The aforementioned equation and approach assume that autofluorescence and background scatter is negligible. The amplitude under those conditions is equal to the inverse of the number of molecules on the membrane ($\langle N \rangle$). If, however, background signal is not negligible, the following modified equation can be used:

$$G(\tau) = \frac{1}{\langle N \rangle}\left(1 - \frac{n_b}{n_s}\right)\left(1 + \Gamma\frac{\tau^\alpha}{w^2}\right)^{-1},$$

where n_b and n_s are the signal from the background and the fluorophore, respectively.

PROTOCOL 9.14: CELL PROLIFERATION USING ANTI-CD71 STAINING (CHAPTERS 4 AND 6)

Cell proliferation is an important metric to determine for many applications. For cell immortalization, it is important to ensure than senescence does not occur. For cell culture in microfluidic devices, it is also important to observe cell growth. Both of these studies can be conducted using cell counting. However, it is often desirable to know how the population is proliferating during the course of growth. Also, in cell-cycle measurements (for cancer studies, embryonic development, etc.), the transition of cells from the G_0 to G_1 phase of the cell cycle is a critical parameter. Likewise, in senescence studies and aging, the removal of cells from the cell cycle to the G_0 phase is an important parameter. While there are several markers of cell proliferation, one of the easiest to use is to monitor the CD71 antigen expression on a cell line. CD71 (human transferrin receptor) is present on proliferating cells and is easily assayed using fluorescent antibodies for CD71. Flow cytometry and microscopy can then be used to determine proliferation. If the techniques are calibrated to absolute antigen density (see Protocols 9.12 and 9.13), then the fluorescence measurements can be converted into antigen density. For cell separations, anti-CD71 conjugated to a surface (for microfluidic work) or a magnetic particle (for MACS sorting) can be used to isolate proliferating cells from senescent ones.

Equipment needed (for fluorescence measurements): Flow cytometer and/or fluorescence microscope with suitable band-pass filters for the fluorescent dye.

Reagents needed: Anti-CD71 coupled to a fluorophore (e.g., anti-CD71 FITC or anti-CD71 PE) (the antibody must match the species of cell line (e.g., mouse anti-human CD71 if human cell lines are used)), buffer (phosphate-buffered saline, pH = 7.4, supplemented with 3% w/v bovine serum albumin) containing $5\,\mu g\,ml^{-1}$ of propidium iodide or other impermeant viability probe.

1. Divide the sample into a stained sample and unstained control.
2. If the cells are adherent, use trypsin to produce a suspension of cells (Chapter 3). Prolonged exposure to trypsin may affect CD71 antigen density. If the cells are already in suspension, proceed directly to step 3.
3. Wash 1 ml of cell suspension (10^5–10^6 cells ml^{-1}) once by centrifugation (1000–1200 × g for 4–5 minutes) and discard the supernatant.
4. Resuspend cells in 100 μl of buffer. Add a suitable amount of antibody for the number of cells tested (typically one test).
5. Incubate at room temperature in the dark for 20 minutes.
6. Centrifuge cells, discard supernatant, and resuspend in 100 μl. Repeat process a total of two times.
7. Discard supernatant on last wash and resuspend in 500 μl of buffer (the volume may have to be increased if the cell concentration is too high for the cytometer or microscope).
8. If the sample is analyzed by microscopy, acquire fluorescence images while avoiding photobleaching. Acquire a set of unstained control images for comparison to background – subtract the stained cells.
9. If analyzing the sample by flow cytometry, measure the unstained sample first, setting the threshold to keep the unstained cells in the first intensity decade (10^0–10^1 on log scale).

    Note: If the instruments have been calibrated with known antigen densities, the same settings must be used each time to ensure accurate results.
10. For data analysis' the median fluorescence intensity of the cells for anti-CD71 staining is an indicator of fluorescence intensity. Cells staining positive for propidium iodide should be excluded as dead cells from the analysis.

9.7   FLUORESCENT LABELS AND FLUOROGENIC PROBES (CHAPTERS 4–7)

Fluorogenic probes, molecules that are nonfluorescent in their intact form, but become fluorescent after a chemical reaction, are useful as sensitive probes for biochemical activity [9,19]. Fluorogenic reagents have been used in other areas of analytical chemistry, for example for on-line derivitization of peptides. This table shows examples of fluorogenic probes based on protein activity in the cell. Most fluorogenic probes for cell analysis consist of a highly fluorescent moiety coupled covalently to a functional group that renders the molecule nonfluorescent. Enzymes in the cell will cleave the functional group so that the dye can then fluoresce (Figure 9.11). In the case of acetoxymonoether (AM) probes, the AM group renders the molecule nonfluorescent and also imparts some permeability to the molecule. One of the most widely used probes for viability is calcein-AM. Once the permeant probe enters the cell, esterases in the cytosol cleave the AM group, producing the highly charged, bright calcein molecule that cannot leave the cell. Like other fluorogenic probes, this allows a buildup of fluorescent product, in many instances enhancing sensitivity (although cytotoxic effects may occur at higher concentrations of probe or product).

Other probes that are not strictly fluorogenic, but do produce a change in wavelength or intensity can also be categorized in this group of compounds. Molecular beacons, hairpin DNA structures that undergo fluorescence quenching in the absence of analyte, can also be used to produce an increase in fluorescence in the presence of an intracellular target or cell surface antigen. For molecular beacons, the hairpin opens up to recognize a specific DNA sequence, disrupting the fluorescence quenching. Other

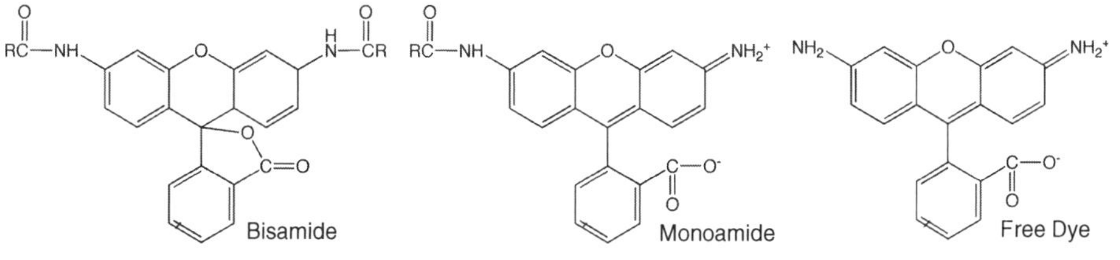

Figure 9.11   Example of a fluorogenic probe. Rhodamine-110, modified with protease substrates (R-groups) as a bis-amide. The nonfluorescent probe is cleaved initially into a monoamide, which is weakly fluorescent (and typically not detected). Cleavage of the second R-group produces the free dye, which is strongly fluorescent

probes that behave in a similar manner include FRET-based probes, where the fluorescent donor and acceptor are linked by an enzyme substrate).

REFERENCES

1. Zhou, S., Lin, J., Du, W. *et al.* (2006) Monitoring of proteinase activation in cell apoptosis by capillary electrophoresis with bioengineered fluorescent probe. *Analytical Chimica Acta*, **569**, 176–181.
2. Southern, P.J. and Berg, P. (1982) Transformation of mammalian cells to antibiotic resistance with bacterial gene under control of the SV40 early region promoter. *Journal of Molecular and Applied Genetics*, **1**, 327–341.
3. Baum, C., Forster, P., Hegewisch-Becker, S., and Harbers, K. (1994) An optimized electroporation protocol applicable to a wide range of cell lines. *Biotechniques*, **17**, 1058–1064.
4. Counter, C.M., Hahn, W.C., Wei, W. *et al.* (1998) Dissociation among *in vitro* telomerase activity, telomere maintenance, and cellular immortalization. *Proceedings of the National Academy of Sciences USA*, **95**, 14723– 14728.
5. Ramirez, R.D., Sheridan, S., Girard, L. *et al.* (2004) Immortalization of human bronchial epithelial cells in the absence of viral oncoproteins. *Cancer Research*, **64**, 9027–9034.
6. Andree, H.A.M., Reutelingsperger, C.P.M., Hauptmann, R. *et al.* (1990) Binding of vascular anticoagulant a (VACa) to planar phospholipid bilayers. *Journal of Biological Chemistry*, **265**, 4923–4928.
7. Trahtemberg, U., Atallah, M., Krispin, A. *et al.* (2007) Calcium, leukocyte cell death and the use of Annexin V: fatal encounters. *Apoptosis*, **12**, 1769–1780.
8. Garcia-Calvo, M., Peterson, E.P., Leiting, B. *et al.* (1998) Inhibition of human caspases by peptide-based and macromolecular inhibitors. *Journal of Biological Chemistry*, **49**, 32608–32613.
9. Hug, H., Los, M., Hirt, W., and Debatin, K.-M. (1999) Rhodamine 110-linked amino acids and peptides as substrates to measure caspase activity upon apoptosis induction in intact cells. *Biochemistry*, **38**, 13906–13911.
10. Peramo, A., Albritton, A., and Matthews, G. (2006) Deposition of patterned glyco-saminoglycans on slianized glass surfaces. *Langmuir*, **22**, 3228–3234.
11. Themo-Pierce Tech Tip #5, Attach and Antibody onto Glass, Silica, or Quartz Surface, http://www.thermo.com/pierce.
12. Hermanson, G.T. (1996) *Bioconjugate Techniques*, Academic Press.
13. Jing, R., Bolshakov, V., and Flavell, A.J. (2007) The tagged microarray marker (TAM) method for high-throughput detection of single nucleotide and indel polymorphisms. *Nature Protocols*, **2**, 168–177.
14. Wang, K., Marshall, M.K., Garza, G., and Pappas, D. (2008) Open-tubular capillary cell affinity chromatography: single and tandem blood cell separation. *Analytical Chemistry*, **80**, 2118–2124.
15. Phillips, J.A., Xu, Y., Xia, Z. *et al.* (2009) Enrichment of cancer cells using aptamers immobilized on a microfluidic channel. *Analytical Chemistry*, **81**, 1033–1039.
16. Shapiro, H.M. (2003) *Practical Flow Cytometry*, 4th edn, Wiley-Liss, New York.

17. Schwille, P.S., Korlach, J., and Webb, W.W. (1999) Fluorescence correlation spectro-scopy with single molecule sensitivity on cell and model membranes. *Cytometry*, **36**, 176–182.
18. Chen, Y., Munteanu, A.C., Huang, Y.-F. *et al.* (2009) Mapping receptor density on live cells by using fluorescence correlation spectroscopy. *Chemistry a European Journal*, **15**, 5327–5336.
19. Ho, N., Weissleder, R., and Tung, C.-H. (2006) Development of water-soluble far-red fluorogenic dyes for enzyme sensing. *Tetrahedron*, **62**, 578–585.

Index

References to tables are given in bold type. References to figures are given in italic type.